Essentials of Pest Management

Essentials of Pest Management

Key Information on Pest Identification and its Management

By

Prakash Rambhat Thalya
Ravi Chandra

WINGS PUBLICATION
WE GIVE WINGS TO YOUR DREAM!

WINGS PUBLICATION

WE GIVE WINGS TO YOUR DREAM!

Registered Office- 907-Sneh Nagar, Sapna Sangeeta Road,
Agrasen Square, Indore – 452001 (M.P.), India

Website: http://www.wingspublication.com

Email: mybook@wingspublication.com

First Published by WINGS PUBLICATION 2022

Title : Essentials of Pest Management

Price : Rs. 699 I $ 14.99

ISBN : 978-93-90661-53-4

Dedication

Its our wish to remember and acknowledge the strength and support we have got from these people without any hue and cry and at all times. We hereby dedicate this book to the following people.

First, to our parents who were our first teachers, philosophers in this world. Second, to our family members without their support we couldn't have come so far, and last but not the least our industry fraternity and colleagues who have contributed in our long journey throughout, rich experience we have gained all these years. We sincerely thank all those mentioned for their support and encouragement.

ACKNOWLEDGEMENTS

We would like to thank our Colleagues, Peers, seniors, Pest Control Industry Fraternity and our Customers who helped directly or indirectly to acquire knowledge, experience and exposure, which allowed us to learn day by day in our combined experience of more than 3 decades. It has helped us to come out with a book, which we feel is going to further continue the chain of giving the same knowledge to the pest control industry fraternity of current and future.

We thank Aparna Shivapura for lending her supporting hand for the content editing, Mr. Srivatsa S and Mr. Suresh for helping in gathering and arranging the insect pictures and all others who have extended their support in making this book a reality and seeing the light at the end of the tunnel.

We sincerely thank and express our gratitude to Mr. KK Unni, Dr. Kedar Deobhankar, Dr. Anil Kumar Makkapati, who have written FOREWORD to this book.

Last but not least, we thank our Publishers M/s Wings Publications, Pune who has spent time in understanding all our needs and supporting the first of its kind book publish and bring this book into the marketplace for the benefit of all the readers.

We thank all people who have patted our backs by lending a few words of appreciation and remarks in their comments for this book which keep us motivated to strive for more and better in the future editions of this book.

PREFACE

INSECTS presence dominate the land fauna with about one million identified species. Entomologists estimate the number could be as high as 5 million. INSECTS live in almost every habitat. Insect fossils a mandible (jaws) dates back to about 400 million years. Insects are very integral and important part of our ecosystem wherein they pollinate plants, produce useful substances and act as scavengers. The perceived notion of all insets are pests is changing and there are efforts now put in the direction of creating favorable bio diversity. Flaura and fauna enabling growth of beneficial insects in many parts of the world.

This book, ESSENTIALS OF PEST MANAGEMENT is an attempt done by authors Prakash Rambhat Thalya and Ravi Chandra to provide insights on the Insect population from the Indian Perspective of Insects that have become pests of Urban India. This book serves as a good ready recknoer on Insect identification, habitat, biology and best practices of management methods. This would be a comprehensive guide to all the professionals of Pest Management Industry, new entrants to industry and Enthusiasts to understand about Insects and urban pest problems and their Management.

TESTIMONIALS

It gives me immense pleasure to announce the launch of "Essentials of Pest Management" a comprehensive ready reckoner on Pests. While there is so much information available on the internet, the arranging of relevant facts on pest control in a concise manner requires in-depth knowledge of the industry. I have had the pleasure of interacting with both the authors Mr Prakash Thalya and Mr Ravi Chandra who have spent enough time with the pest control industry to know the exact requirements and present it through this book. I thank the authors on behalf of the Indian Pest Control Association for contributing back to the industry from which they have learnt their skills through their book. The combined experience of both the authors, which comes with practical on the field experience will be the hallmark of this book. I am sure the book through the knowledge it imparts, will help in establishing a new knowledge order in the pest control industry and will remain continuously relevant in the times to come

Prakash Sasidharan
President
Indian Pest Control Association

Indian Pest Management Industry is one of the largest in the world led by few established progressively and other individuals and few startup companies. It is observed that there is scarcity of new educational material for their day to day upgradation of knowledge. There was always a need of some ready reckoner material in this regard. The book Essentials of Pest Management having focus on pest identification, techniques of management will be apt book for the industry. I congratulate Mr. Prakash Thalya and Mr. Ravi Chandra for writing this book and I am sure our industry will certainly get benefited from this book. My best wishes to both the Co-Authors for the better success.

Dr.Sarang Savalekar
President
Pest Management Association of India
Director, Paramount Pest Consultants Pvt. Ltd.,

I am very happy and exuberant to give my comments for the book – Essentials of Pest Management authored by Mr. Prakash Thalya and Ravi Chandra. This is a wonderful thought and effort by them to bring out such a unique book for our Industry and for the first time in India as per my knowledge.

Prakash and Ravi are professionals of pest management industry having vast experience and have utilized the same in bringing this book which will eventually become a hand book of every Pest Management Professional.

I urge all the pest control companies and people associated with Pest Management to take the benefit of this very well written book. I wish a grand success to Prakash and Ravi in their endeavor.
Balamurali KR
President
Tamil Nadu Pest Management Association

Congratulations Mr. Prakash Thalya & Mr. Ravi Chandra on authoring a book on Pest Management. This is an incredible milestone and need of the hour for every Pest Management Professional, so you deserve the spotlight to celebrate the moment. Such Articles writing for a book can feel daunting but as you break down the process into steps, you must have realized that writing industry articles is like writing for any other genre. Your ability to relentlessly search for solutions to problems and find innovative ways to improve the industry is the key to this success and many to come.

I have gone through this sample book. You seem to have covered important topics viz: Insect name, its family, Physical appearance, Biology, Habits Control measures of each and every insect. Your book is very interesting where one can pick up important tips which can be implemented to improve our business. It is key that all knowledge and experience is passed on and shared within the industry. So, Cheers to you for a job well done! Because of your creativity and passion, it's no surprise that your book will become very successful.

Raju Parulkar,
Former president, IPCA.

At the outset I am very much pleased and happy with the contents covered in this book.

Insects are the largest Group in Animal kingdom both types and numbers. Apart from Beneficial insects there are large number of insects which eat, damage and destroy food. Some bite, suck blood act as vectors responsible for diseases and death e.g. Mosquitoes and fleas. Many others contaminate food with disease causing germs rendering large number of diseases in the process e.g. Flies and cockroaches etc. Apart from this Termites and fabric pest cause colossal damage to property. So, insects affect Health Food and Property which is dear to Mankind, Identification of the pest is most important step before deciding correct strategy for its control or face a failure, damage and loss of reputation.

The Authors have decades of close association with the subject in the field covering India and neighboring countries. I appreciate efforts put in this publication which is simple and easy to understand. This will help all the Pest Control Operators as standard reference book to correctly identify the pest in the field and control strategy to be used.

Suggest them to come out with more publications in this field to help the industry

All The Best for their Maiden Effort coming out with user friendly publication.

Laxmikant Naik
Industry veteran & Consultant

It is a privilege to give my comments and appreciation to Mr. Prakash Thalya and Ravi Chandra for their efforts to bring out a unique first time in our industry a self-reckoner tool in the form their book Essentials of Pest Management. Prakash and Ravi are professionals of pest management industry have written such an elaborate book on the subject. Prakash and Ravi from their long professional experiences have captured meticulously information which will make pest controlling and managing most effective. I urge all the pest controlling companies and people associated with Pest Management to take the best benefit of this very well written book. Their endeavor is very well appreciated and I wish them all the best in this new endeavor.

Dinesh Bapat
Consultant – Sumitomo Chemicals – HHI segment

"Essentials of Pest Management comes at a right time when Health & Hygiene is considered very important. I congratulate the authors for taking these extra ordinary efforts and this is a very significant contribution to the Indian Pest Control Industry. Indian Pest management Professionals are now equipped with the right information about understanding and management of each and every pest and it is a must read for anyone who are interested to manage the pest challenges in a scientific way"

M. Arun Kumar
Head – South Asia
Bayer Environmental Science

I know both Mr.Prakash Thalya and Mr.Ravi Chandra since 2002 when I have decided to step in to public health pest control business. You both have deep knowledge of majority pests, with your experience in understanding and managing through different ways of pest management. Indian pest control industry didn't have any such hand book regarding identification of pests. This book Essentials of Pest Management will really help new comers in pest control business and also training purposes for pest control technicians of existing pest control organizations. I wish both authors all the best for their new endeavor.

Mukesh Patel
Managing Director
Arbuda Agrochemicals Pvt Ltd.,

"Herewith grabbing the opportunity to comment and appreciate Mr. Prakash Thalya and Mr. Ravi Chandra as we have witnessed their long and effective journey in the Pest Management industry. Their effort to bring out, for the first time, a unique, self-reckoning tool in the form of their book "Essentials of Pest Management" will be a big helping hand in the industry.

Mr. Prakash and Mr. Ravi are professionals of the pest management industry, which resulted in such an elaborate book on the subject, from their long professional experiences, have captured meticulously information which will make pest controlling and managing most effective. I urge all the pest controlling companies and people associated with Pest Management to take the best benefit from this very well written book. Their endeavour is very well appreciated and we wish them all the best in this new endeavour."

Jagdish Shah
Director
A. Saj Agricare Pvt. Ltd.

My congratulations to writers for bringing this excellent book. The decades of experience of authors are truly reflected in this book, which will be useful guide to all interested in the field of pest management in urban areas. This includes not only the Pest Control Operators but also those who manage large buildings, like hotels, malls, office complexes, housing societies and even individual house owners.

The contents are neatly documented with easy to find answers for all questions on Pest Management of any pest starting from its biology, habitat and Management. Very clear pictures are added, which completes the wish list of a curious reader. The control measure includes the names of products which can be used. This will become a Ready Reckoner for Pest Control Industry.

Anil Jain, Executive Director
Crystal Crop Protection Limited

A passion to give the best by the professionals involved in pest control and management resulted into such an elaborate book on the subject. Prakash and Ravi from their long professional experiences learnt what would work best and thus captured every meticulous information which will make pest controlling and managing most effective. I will urge all the pest controlling authorities to take the best benefit of this very well written book. The endeavor is very well appreciated.

Prashanta R Das, VP, Sales and Marketing,
Nuziveedu Seeds ltd, Kolkata

Every person from their childhood knows about pest like cockroach, mosquito, rat etc. Their knowledge is limited to names of the pest. To manage these pests, we have our pest control professional (PCOs). It is very important for PCOs to know the pest and its impact on human life, so that he/she can give a professional service to their customers. Knowledge is essential for one's growth. Pest Control is a specialized field and it requires continuous upskilling. This book is authored by industry veterans with vast knowledge in pest control. It is an excellent handbook for PCOs to Know Your Pest (KYP). Most of the household pests that are covered in this book are seen across our Indian sub-continent. Uniqueness of this book is, it covers the pests with their biology and management in simple language for a common man's understanding. I am sure this book will become a bible for Pest Control Professional.

Anil Abraham
Chief Operating Officer
KAPL, Kottayam
Former Head of Marketing, Bayer – ES division, India

"Knowing your adversary well, helps plan and execute your winning strategy.

This handbook is well thought and designed to equip the reader with intricate knowledge on the identification, pest habitat and dynamics to effectively manage the pest. It's an important tool for PCO & warehouse managers to nip the problem in initial stage, saving precious living resources and avoiding wastage of food grains. The authors with their practical experience of the ground level challenges faced in tackling these pests, have put together a valuable treasure chest of techniques which can be adopted easily.

Chaitanya Dhareshwar
Agri. Solutions Professional
Ex Hoechst, Aventis, Bayer CropScience Ltd

Foreword 1

It is my privilege to write a Forward to "Essentials of Pest Management" authored by my colleague Mr. Prakash Thalya and a veteran Pest Management professional, Mr. Ravi Chandra. Incidentally I was actively associated with pest management – household – and outside environment pest control in Hoechst, AgrEvo, Aventis Cropscience, and Bayer Crop science days.

Right from the 1950s chemical pesticides have been used mainly in agricultural crops in India. Crop pest control including fungi was shifting from inorganic chemicals to organic chemicals. Much later came the herbicides – selective and non-selective. Another focus area was control of locust's swarms of which devastate everything green in North/ Western India.

The war between man and mosquito would continue – neither of the two will win. One major pest is controlled with continues efforts; a minor pest assumes the role of the major causing severe damages. There are several examples of these phenomena in the Crop Pest management field.

Household pests came into focus much later, when urbanisation assumed significance. The dominant segment were cockroaches, flying insects and rodents. Same time the stored grain pests started causing severe damage in warehouses resulting in the need to focus on prevention and control methods in this segment separately.

There are few scientific reference books so far on these household and outdoor pests and controlling them. A serious attempt is made by Prakash & Ravi Chandra in compiling a valuable reference book "Essentials of Pest Management "with an intention to support pest management professionals in India. This book is very well written in a professional manner as Guide Book covering 64 pests – identification, habitat and suggested Management measures. There are attractive colour pictures of all these pests for easy identification.

I am sure "Essentials of Pest Management" will greatly help current professional pest control operators and provide valuable scientific information the new entrants in this segment of business.

I compliment the authors – two veterans with over 30 years of experience

to have thought of compiling this exhaustive book – a much needed Hand Book. I am sure many engaged in this business segment will benefit immensely. I do hope that Ravi Chandra and Prakash will update the technology changes and changes in pest behaviour as and when the 2nd edition will materialise.

Kannan K. Unni
(Ex-Chairman & Managing Director – A Bayer Group Company)

Foreword 2

It is a pleasure to write a foreword to this excellent book brought out by two authors- Prakash Thalya and Ravi Chandra who have decades of experience in Pest management.

Control of pests is important in preventing transmission of diseases to mankind and livestock, ameliorating the nuisance caused by their bites and also preventing structural damage to premises (termites) and food (rodents and stored grain pests).

With onset of global warming, increased urbanization and mobility, a conducive environment is created for variety of pests to thrive among human habitats. This calls for greater understanding of pest biology and their role as vectors or nuisance pests and the control measures advocated.

Many pest control operators are faced with variety of problems concerning how to manage certain pest infestations. To effectively control any particular pest infestation knowledge of pests, their biology and habits and choice of right control methods are essential.

The authors have splendidly elaborated identification, habits and Management methodologies covering wide range of insects from ants to termites including avian and stored grain pest in a very lucid manner. Products recommended for control of various pests, their dosages and application rates have been elaborated in detail.

There are many books that have been written on biology of insects and their management but this book covers information on broad range of pests and their control and offers practical solutions to control of pests. This book would serve as useful reference guide to students in entomology and pest control operators which would help them in instituting appropriate control measures.

I strongly recommend this book for the pest control operators and if translated into local languages would benefit the pest control community at large.

Kedar Deobhankar, Ph.D
Managing Director
Ross Lifescience. Pune

Foreword 3

Climate change, rapid urbanization, deforestation, people migration, improper water storage and supply are some of the reasons for pest explosion all across the globe. We have problem of increase of density of pests on one hand and invasion of pests to newer areas on the other hand. Adding to this is the growing concern of pesticide tolerance, resistance in pests.

Growing urge in the most city dwellers especially millennials, to lead lives in good hygienic conditions (pest free environment to say) demands of quality pest management services. Quality pest management services means, right products usage, with appropriate knowledge about pests and their biology coupled with understanding of local regulatory requirements. Adding to this would also be the great skills at customer relations. Like in many professional businesses, pest management professionals also require to update, refresh their knowledge on existing and new pests, pesticides, pesticide applicators, regulatory guidelines, compliances, etc.

Pest management is not just Science. It is both science and art. Without proper knowledge about pests, even pest pesticide would be redundant.

The handbook “Essentials of Pest Management” nicely written by Ravi Chandra and Prakash Rambhat Thalya fills the gap of authentic and lucid knowledge resource for pest management professionals and likes especially in India. The book clearly showcases the vast techno-commercial knowledge acquired by the authors over the past three decades. I have the experience of working closely with both the authors for over decade and have huge respect for their commitment and passion they have their work and industry.

In this book, the authors put together knowledge in a manner which can be understood easily for newcomer into the industry. Congratulations to the authors for their efforts and good wishes to all readers of the handbook.

Anil Kumar Makkapati
MSc., (Medical Entomology), PhD., FISCD

TABLE OF CONTENTS

1.

Aedes Mosquito

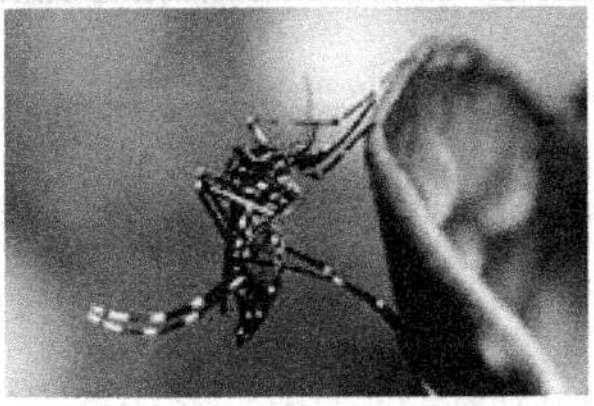

Order / family: Diptera/ Culicidae
Scientific Name: Aedes spp.

Description:

Mosquitoes are members of the order Diptera, i.e., flies. Mosquitoes are slender, long legged, two-winged insects that are 1/8 to ¼ inch long. They are unlike other flies that their wings and bodies are covered with scales, and have long piercing and sucking mouthparts.

The larvae, or wrigglers, are aquatic and have well developed heads, swollen, unsegmented thorax, and eight-segmented abdomens, which end in an elongated breathing tube. The pupae are also aquatic and active. Their heads and thoraxes are encompassed in a large, oval mass with the slender abdomen attached. The most notable species are the yellow fever mosquito (A.*aegypti*), Asian tiger mosquito (A. *albopictus*), saltmarsh mosquito (A. *sollicitans*), and inland floodwater mosquito (A. *vexans*). The adults of these species are described below:

The yellow fever mosquito has silvery white or yellow-white stripes on its black body and a pattern of stripes on the top of its thorax (segments with legs and wings attached) which looks very much like a tuning fork. The last five segments of the legs are banded in white.

The Asian Tiger mosquito has silvery-white markings on its black body; the top of the thorax has a single silver-white stripe down the centre and silver-white bands around the abdominal segments. The last five segments of the legs are banded in white.

The salt marsh mosquito has white bands on the last leg segments and a band across the middle of the mouthparts. The abdomen has bands on most of the segments and a white stripe down the middle of the top of the abdomen.

The inland floodwater mosquito is medium sized and brown with a bad of white scales at the end of each abdominal segment. Each band is notched with a dark "V" that appears to cut it into two pieces. It has very narrow, creamy white to bronze bands on the last five segments of the legs.

Biology:

The eggs, which are laid singly above the water line, hatch when the water rises and contacts the eggs. In the absence of sufficient water, the eggs remain dormant for many years. Mosquitoes undergo complete metamorphosis, i.e., egg, larva, pupa, and adult. The larva and pupa require water for development. Under ideal conditions, these mosquitoes complete their life cycles in ten days.

Common Site of Occurrence:

Aedes Mosquitoes are normally a Day biting insect it would be active in Day time. They breed in the artificial or man made water collections like flower pots, thrown away plastic cups heap, discarded or unused material debris, thrown away tyres, tender coconut shells, construction activities holding water stagnation, uncovered stored water containers, uncovered overhead water tanks. Indoors they rest in a dark areas close proximity of humans.

Habitat and Economic Importance:

Asian tiger and yellow-fever mosquito larvae are found in artificial containers, such as discarded tin cans, tyres, flower containers, in cemeteries, etc. Both species are daytime biters, although the yellow-fever mosquito also bites at dawn and dusk, as do most other species of mosquitoes. In summer, mosquitoes become major nuisances as they seek out your customers for their blood meal. Most service calls result when a customer plans activities in the backyard, such as an evening party, or there is a problem at recreational sites located in inaccessible areas of the customer's premises, such as swamps, storage sites, and in other sites which are difficult to manage, such as city sewers and under structures.

Management:

The most common methods of collecting mosquitoes for identification are the use of light traps, siphoning tubes at resting sites, and dippers to collect larvae. Both styles are designed to catch specimens without destroying them on sticky boards or by electrocution. Adults can be collected by using a siphoning jar trap. The best areas for collection during the day are in shady areas, such as inside and on the shady side of sheds and buildings, sewers, culverts, drainage pipes, etc.

Lighting around the property can result in problems with mosquitoes. Because lights are beacons for mosquitoes, attraction to the area should be reduced by locating lives away from the structure and directing the mosquitoes towards them. Sodium vapour and/or yellow bug lights should be used instead of incandescent or mercury vapour lights. Regardless of the species, the key to mosquito control is source reduction, i.e., the removal of the water breeding sites. For container breeders, this is usually resolved by site clean-up. The two most popular larvicides are the insect growth regulator (IGR), Methoprene and the bacteria Bti. Liquid, granules and extended-release formulations are available to treat the site even water are not present. When the product is exposed to water, the active ingredient is released into the water and kills larvae.

Aerosol and ultra-low volume products are effective in killing the mosquitoes within the immediate area but does not provide long term control. Residual spraying of mosquito resting sites also reduces the problem. The use of repellents should be considered. Skin repellents, such as DEET and clothing repellents that contain permethrin, are very effective. Mosquito problems require considerable customer education because of the diversity of species, their often-remote breeding sites, and the ability of many species to travel long distances.

2.

Almond Moth / Tropical Warehouse Moth

Order: Lepidoptera

Family: Phycitidae

Scientific Name: Ephestia cautella Walker or Cadracautella

Description:

The Almond Moth or Tropical Warehouse Moth is a small, stored-product pest. Almond moths feed on cereals, oil seeds, groundnuts, palm kernels, cocoa, spices, animal feeding stuff, flour materials, garlic, dry chillies, dried fruits, etc. A primary pest, webbing and frass are produced in the infested product. This species may be confused with the related Indian Meal Moth (*Plodia interpunctella*) or the Mediterranean Flour Moth (*Ephestia kuehniella*), which are also common pantry pests in the same subfamily. Other common names are Dried

Currant Moth and Fig Moth. The Almond Moth has achieved a cosmopolitan distribution due to transport with food products in its larval form.

Biology:

Adults live for about 10 days after eclosion and do not eat but may drink water, if available. Adult Almond Moths are predominantly light brown in colour; forewings are dull greyish in colour, smaller grey hind wings that are typically grey, while the back edges of the wings have darker bands at the edges. Larva: Eruciform, have setae arising from pigmented spots on the cuticle, look brownish-white in colour, larval head dark brown.

Habitat and Economic Importance:

Here, the infestation is confined to the surface layers of the stored commodity. The adults show a very marked periodicity in their flight activity and egg laying; their active flight takes place from 5 to 7 pm and at 6 to 7 am when fluctuations in temperature and humidity occur. The adults avoid strong light and rest in dark places during the daytime. Nearly 300 eggs are laid on the grain material and in gunny bags filled with grain. A maximum number of eggs are laid within 3 to 4 days of emergence and hatch in 3 days. The larvae move freely in the food material, contaminating it with webbing and frass material. The larvae feed largely on the embryo of the cereals. Before pupation, the larvae enter a wandering phase and completely obscure the bag surfaces with silken webbing. The life cycle takes about 27 to 31 days under optimum conditions.

Common Site of Occurrence:

These insects are mainly seen in the storage houses in the field, godowns, warehouses, and if there is no proper fumigation control methods it will be appeared in the packaged or finished product goods sections as well. The type of feed which is destroyed by these insects are Cereals, oil seeds including groundnuts and palm kernels, cocoa, spices, animal feeding stuffs, flour materials, garlic, dry chillies, dried fruits, etc.

Management:

Firstly, one has to clear all food items where infestation occurred. A vacuuming is required to be done thoroughly in all areas of stored food to ensure all eggs, larvae, adults and webbing are removed. Pheromone traps can be used to identify the areas of activity if this is not clearly known. Insecticidal residual sprays can be applied on surfaces where storage is going to be done with Synthetic pyrethroids. Ensure all cracks and crevices are treated thoroughly. Infestations in grain storage or grain processing facilities warehouses can be controlled using intensive sanitation and fumigation techniques.

3.

American Cockroach

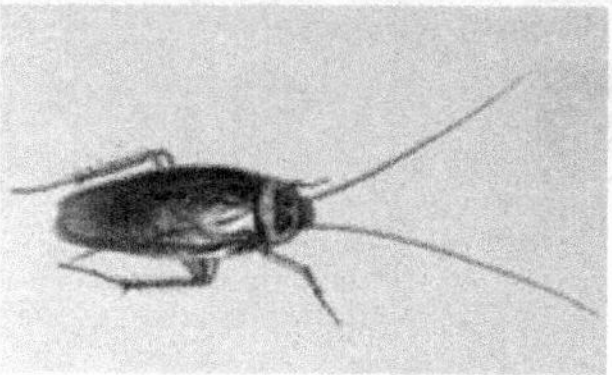

Order/Family: Blattodea/ Blattidae

Scientific Name: Periplaneta americana (Linnaeus)

Description:

American cockroaches are 1-3/8 to 2-1/8 inches long when mature, red-brown, and characterized by fully developed wings that completely cover the abdomen. The pronotum (i.e., shield like area behind the head) has a yellow band around its edge.

The nymphs are quarter-inches long when they emerge from their capsule. As they develop, they become more red, brown and the yellow band becomes more prominent on the pronotum. The purse shaped egg capsule (i.e., ootheca) is dark red-brown in colour 3/8 inch long and typically as eggs per side.

Biology:

During her lifetime, the female American cockroach produces from 9 to 10 egg capsules, each of which contains from 14 to 16 eggs. The capsules are dropped or, using secretions from her mouth, glued in protected locations such as cracks and crevices near food sources. The nymphs molt from 10 to 13 times before becoming adults. This requires about 600 days. Adult females live an average of 440 days and males about 200 days. Large populations of American cockroaches accumulate in secluded locations because they have a long lifespan.

Habitat and Economic Importance:

American cockroaches are common pests in most homes. They can be abundant in sewers and commercial facilities, including grocery stores, prisons, restaurants, hospitals, office and apartment buildings. They prefer to inhabit warm, damp locations, like steam tunnels and boiler rooms. They fly and easily migrate from building to building. In the summer, large numbers accumulate in outdoor locations, in dumps, alleys, and migrate into surrounding structures. Mass Migration is one of the characteristic features of American Cockroach which results in sudden up swing of infestation

Common Site of Occurrence:

These insects are water loving insects and are found in near vicinity of water like bathrooms, sewage lines, sewage manholes, kitchen sinks, storage areas, packing material storages, garbage dumps etc places in households, offices, eateries, retails, warehouses, stores, etc commercial establishments.

Management:

Cockroaches often are brought into the facilities via equipment and storage boxes. Thus, potentially infested products that are bought into structures should be closely inspected. Cockroach monitoring traps are available to help pinpoint sources of cockroach infestation and to monitor areas, but infestations cannot be visually detected. Good flushing agent treatments like propoxur can be used for spray treatment to flush cockroaches from their harbourages. An effective cockroach management program depends on good sanitation in order to eliminate the food, water and harbourage they need for survival. It is critical to reduce clutter as large cockroaches hide in stacked boxes, cartoons, rolled carpeting and stored paper and cardboard materials, especially in the dark, damp locations. Vacuum cleaning can be used to physically remove exposed cockroaches& their egg case.

American cockroaches are particularly sensitive to drying, so it is important to reduce moisture by repairing leaks, improving drainage, and installing screened vents in order to increase airflow. The most important cracks at which sinks and fixtures are mounted to the wall and/or floor, around all types of plumbing, baseboard moulding; cabinet meet walls and door frames, cracks on or near food preparation surfaces. Cockroach access routes from wall voids into occupied spaces and around plumbing and electrical fixtures should be sealed with caulk or grout. Basement floor drains should be protected with screens or basket inserts which must be cleaned regularly. Containerized, paste, dry and Gel baits have become very popular in the industry and are very effective products for eliminating cockroaches. Large, plastic bait containers should be placed as close as possible to the dark concealed spots where cockroaches are living, preferably adjacent to edges and corners.

A large population of cockroaches can be reduced or eliminated with careful application of sprays and dusts, followed by application of paste, gel and dry baits. Many insecticides break down rapidly within moist, hot locations which these cockroaches prefer, so an appropriate formulation for the environmental conditions must be selected.

4.

Angoumois Grain Moth / Paddy Moth

Order: Lepidoptera

Family: Gelechiidae

Scientific Name: Sitotroga cerealella

Description:

Readily distinguished from all other moths by its small size, light straw or buff colour, the fore wings are trapezoidal with fewer darker markings. The hind wings abruptly narrowed at the apical end. The posterior margins are heavily clothed with long hair (setae). The hind wings are gray and taper to a thumb-like projection at their tips. Matured larvae about 1/2" long and yellow-white with brown heads. Thorax is much stouter than the abdomen, which is gradually narrowed behind. Pupa, typically object type, pupates within the kernel. Products infested include corn and wheat, paddy, barley, beans, cowpeas and sorghum. It is a primary pest easily killed by cold winters and is a serious pest in the tropics.

Biology:

Adult moths lay their eggs in the gaps between kernels. Upon hatching, the tiny caterpillars immediately bore into a grain kernel, covering the hole after they enter. These larvae go through three Moults while feeding on the interior of the kernel. Just before pupating, the mature caterpillar chews a circular hole leaving a thin film over the future exit, through which the adult moth emerges. The length of this life cycle depends on temperature, with completion in 30 days at 300C and 40 days at 250C. Generally, they will have four to five generations per year, with the larvae being dormant in the coldest winter months; but may have as many as 10-12 generations in heated warehouses.

Habitat and Economic Importance:

It is European in origin, first reported from Angoumois, a French province. Moths prefer damp grains infest the grain in the field itself. Standing crop may be attacked, but a very severe outbreak occurs in the grain harvested but not threshed and stored. Infestation in stored grain is largely restricted to the periphery. The females lay up to 200 eggs during the life span of 5 to 10 days. The newly laid eggs are pearly white and remain so if they are unfertile. The fertile eggs become bright red or pink, usually within 24 hours of incubation. The eggs are deeply fluted longitudinally. The groves are especially pronounced at the smaller anterior end. The larva gnaws its way out of the egg shell from its smaller end. Upon hatching, the larva is white or buff-pink translucent and bores into the kernel and close the entrance hole with a spun silken web, and feeds upon the soft endosperm of the kernel, forms a cocoon inside the kernel just below the seed coat with a weakly fastened flap or doorway, pushing which the adult emerges out of the kernel. The life cycle takes nearly 4 weeks at the optimum conditions of 300C temperature and 75% RH. In colder climates, the life cycle may take 4 to 6 months, as the larvae remain dormant.

Common Site of Occurrence:

These insects are mainly seen in the storage houses in the field, godowns, warehouses, and other paddy storage areas. If left unattended and in most conditions, this can also be seen in our household grain storage also. The type of feed which is destroyed by these insects are Corn and wheat, paddy, also occurs on barley, beans, cowpeas and sorghum.

Management:

Angoumios Grain Moth (AGM) is the only stored product insect known to infest standing grain in the field. Studies indicate infestation is reduced if grain moisture is greater than 30%. Below that level, the pest prefers damp grain in preference to very dry grain. Carry-over grain can be a major source of AGM and other stored grain insect pests. It should be removed and discarded (sold, fed, buried, or spread on open ground in winter) to prevent the emerging adult moths from flying back to re-infest the newly stored grain. If leftover grain must be retained, it should be properly fumigated with appropriate fumigation techniques.

Removal of all spilled grain from in and around the storage area is an important method of reducing the infesting population. All equipment used to harvest, move, and store the grain should be thoroughly cleaned well before these events occur. Insecticide application of Pyrethroids space spray can be applied for a quick knockdown in the high infested zone.

5.

Anopheles Mosquito

Order/ Family: Diptera / Culicidae

Scientific Name: Anopheles spp.

Description:

Mosquitoes are members of the order Diptera, i.e., flies. Mosquitoes are slender, long legged, two-winged insects that are 1/8 to ¼ inch long. They are unlike other flies as their wings and bodies are covered with scales and they have long piercing and sucking mouthparts. The Anopheles Mosquitoes have no scales on their abdomens, as do Aedes or Culex mosquitoes. They are easily recognized by their mouthparts (i.e., two palps and proboscis) which are equal in length, as well as their habit of resting in a straight line at a 45° angle to the surface.

The larvae, or wrigglers, are aquatic with well-developed heads, swollen and unsegmented thoraxes, and eight-segmented abdomens, which usually end in an elongated breathing tube. *Anopheles* larvae have no elongated breathing tube as do the Andes or Culex mosquitoes, and they have orient themselves parallel to the surface of the water when they are resting or breathing. The pupae, or tumblers, are also aquatic and active. Their heads and thoraxes are encompassed in a large, oval mass with a slender abdomen attached.

Six Anopheles species, Anopheles baimaii, Anopheles culicifacies, Anopheles fluviatilis, Anopheles minimus, Anopheles stephensi, and Anopheles sundaicus are implicated as primary vectors transmitting malaria in different eco-geographical regions of India. Anopheles culicifacies is the major species of malaria transmission mosquito in India. The adults of these species are dark brown, legs dark or with some pale patches, and have dark scales on each wing. They are medium sized, brown mosquitoes with hairy, second body segments which are white in the middle and dark brown on the sides. Anopheles stephensi is another species prevalent in India which has emerged as an urban malaria vector

Biology:

Anopheles mosquitoes form swarms during dusk for mating. Both male and female mosquito feed on nectar for energy. It is only after mating and for the development of the egg's female mosquitoes feed on blood. For some species one feed of blood is sufficient to lay eggs and for some species second blood feed is required before laying eggs. After the blood meal they stay secluded until they are ready to lay eggs. They lay eggs, which have tiny floats attached to them, singly on the surface of clean, unpolluted water. The eggs hatch in two to six days, with total development taking about two weeks. Under ideal conditions duration of egg to adult may vary 7 days at 31 degree Celsius and 20 days at 20 degree Celsius temperature. Longevity of Anopheles mosquito varies with the temperature and humidity factors. The average lifespan of female anopheles mosquito is about 15 days but there are instances where the lifespan of up to two months reported.

Habitat and Economic Importance:

Anopheles culicifacies is common in India accounting to about 70% of malaria cases in India. The adult mosquitoes breed in ditches, ponds, swamps, and puddles. These mosquitoes are problems in areas where extensive irrigation of crops occurs. They enter structures to feed on humans, but although aggressive, they are not painful biters. They prefer to breed in the sunlit water stagnations.

Anopheles stephensi typical breeding sites are artificial containers, such as domestic wells, cisterns, overhead water tanks, and roof gutters, with clean and unpolluted water allowing rapid expansion to urban environments. The introduction of Anopheles stephensi exhibits resistance to several classes of insecticides, posing challenges in controlling the spread to new areas.

In the summer, mosquitoes become a major nuisance as they seek out humans for blood meals. Most service calls result when a customer plans activities in the backyard, such as an evening party, or there is a problem at the recreational sites. Perennial problems result from mosquito breeding sites located at inaccessible areas of the customer's premises such as salt marshes, swamps, storage sites, and in other difficult to manage sites, such as city sewers and under structures.

Common Site of Occurrence:

During the rainy season, rain pools and water bodies created by the seasonal Gash River serve as the main breeding sites. In the dry season, irrigation canals, seepage from water pipes, neglected wells, artificial containers, and man-made ditches serve as the main breeding sites. Male adult mosquitoes hide and rest in the shrubs and bushes. Female adult mosquitoes rest in the dark corners and on the vertical objects nearer to the blood source in the households, cattle sheds, indoors in commercial establishments like offices, schools, cinema halls, human transportation vehicles like bus, trains, hospitals etc places.

Management:

It is important to identify the species which is causing the problems as well as the breeding sites. The most common methods of collecting mosquitoes for identification are the use of light traps, siphoning tubes at resting sites, dippers to collect larvae. Light traps can be small and portable or large and permanently wired. Both styles are designed to catch specimens without destroying them on sticky boards or by electrocution. Adults can be collected by using a siphoning jar trap. The best areas for collection during the date are in shady areas, such as inside and on the shady side of sheds and buildings, sewers, culverts, drainage pipes, etc.

Collection of larvae is more labour-intensive because it involves investigation of potential breeding sites which may not be accessible, as well as sampling the site in several locations using a dipper.

Lighting around the property can result in problems with mosquitoes. Because lights are beacons for mosquitoes, attraction to the area should be reduced by locating lives away from the structure and directing the mosquitoes towards them. Sodium vapour and/or yellow bug lights should be used instead of incandescent or mercury vapour lights.

Regardless of the species, the key to mosquito control is source reduction, i.e., the removal of the water breeding sites. For container breeders, this is usually resolved by site clean-up. However, flood water mosquito problems usually require the use of larvicide. The two most popular larvicides are the insect growth regulator (IGR), Methoprene and the bacteria Bti. Liquid, granules and extended-release formulations are available to treat the site even water are not present. When the product is exposed to water, the active ingredient is released into the water and kills larvae.

Adulticiding with Spraying, aerosol and ultra-low volume products is effective in killing the mosquitoes within the immediate area but does not provide long term control. Residual spraying of mosquito resting sites also reduces the problem. The use of repellents should be considered. Skin repellents, such as DEET and clothing repellents that contain permethrin, are very effective. At their best citronella candles and similar products provide temporary relief. Mosquito problems require considerable customer education because of the diversity of species, their often-remote breeding sites, and the ability of many species to travel long distances.

6.

Bats

Order/ family: Chiroptera / various

Scientific Name: Eptesicus fuscus – Big Brown Bat.

Myotis lucigugus – little brown bat.

Description:

Adult species of Bats are 2 ¼ to 7 ½ inches long with a wingspread of 6 to 15 inches. They are tan to black, have large ears, with hair-covered bodies.

Biology:

Nursery colonies typically are formed in locations other than the overwintering site. Depending on the species, one or two babies are born from May through July. The young hang in the roost or are carried by the female. They are weaned and able to fly within a month and leave the roost by the end of August. The average lifespan of most batches is 4 to 10 years; Some, however, lives as long as 19 to 38 years.

Habitat and Economic Importance:

Bats consume thousands of insects per night; humans fear them due to superstition as well as the potential health threat to humans. Rabies and histoplasmosis occasionally are of concern when dealing with a bat infestation in structures. In most colonies, only a small percentage of the bat population is infected with rabies, and because Bats avoid contact with humans and domestic animals, the incidence of human exposure to Bat rabies is low as bat droppings accumulate in the roost, the risk of histoplasmosis increases.

Common Site of Occurrence:

Bats being a nocturnal mammal during the day time they rest on trees, rock crevices, caves, building shafts, building high rise structures. They do fly to a steam, pond or lake in the vicinity to dip their lower jaw into water and drink water. In buildings and household locations they use wall attic spaces for their resting and roosting place.

Management:

Inspection to determine where the bats are entering the structure should be conducted approximately ½ hour before dusk until one hour later. Gaps around vents, chimneys, roof lines and other areas are potential areas for Bats to be found. The attic area should be inspected to determine the size of the colony, other points of entry, the number of accumulated droppings, and the need for foot dropping removal and disinfection.

The best long-term solution for bat infestations should be used only from September through April after the bats have migrated or when dealing with non-migratory species in order to avoid trapping newborn bats within the structure. All entry points greater than 3/8-inch wide should be sealed. The most effective procedure is to seal all but one or two openings, thereby allowing three to four days for the bats to adjust and then to seal the remaining holes one evening after they emerge from the structure.

Using mothballs and light in order to repel bats is of limited value because Bats typically relocate within the present structure. If a stray bat enters the interior office structure, it usually can be encouraged to leave through an open door or window. It also could be captured, e.g., under a coffee using a piece of cardboard as a lid and releasing the bat unharmed outdoors. Under no circumstances should the Bat be handled.

A residual insecticide and/ or ULV application should be made to the roost area in order to control ectoparasites, e.g., bat bugs and mites. Excessive dropping accumulation should be removed and the area disinfected.

7.
Bed Bug

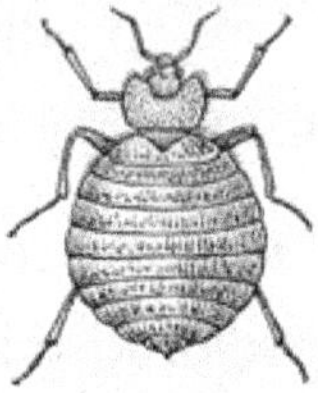

Order / Family: Heteroptera / Cimicidae
Scientific Name: Cimex Lectularius (Cimex Sps)

Description:

The adult Bed Bug is 3/16 inch long, oval, flat and rusty-red or mahogany in colour. It is flat and thin when unfed but becomes more elongate, plump and red when it is full of blood. Four segmented antennae are attached to the head between the prominent compound eyes. The three segmented beaks, or proboscis, is located beneath the head and passes back between the front legs. This insect cannot fly as its wings are reduced to short wing pads.

Biology:

The female bed bug lays eggs, i.e., one to five per day and 200 to 500 within a lifetime. A clear substance gets attached to cracks and rough surfaces. Under ideal conditions, eggs hatch in about seven days and the nymphs molt five times, taking a blood meal between each molt. Development time from egg to adult is three weeks / 21 days. Adults can live for almost a year.

Habitat and Economic Importance:

The Bed Bug hides in cracks and crevices during the day, preferring to rest on wood and paper surfaces instead of stone and plaster. It leaves these harborage areas at night to feed on its host, which includes humans, birds, and family pets. The blood meal requires three to ten minutes and usually goes unnoticed by the victim. After feeding, the bite site may become inflamed and itch severely in sensitive people. Although it has been associated with over 25 diseases, transmission has not been conclusively proven. Over time, the harborage areas become filled with the molted skins, feces and old eggshells of the resident bed bugs. These areas have a characteristic smell caused by a secretion emitted by the Bed Bug.

Common Site of Occurrence:

Bats being a nocturnal mammal during the day time they rest on trees, rock crevices, caves, building shafts, building high rise structures. They do fly to a steam, pond or lake in the vicinity to dip their lower jaw into water and drink water. In buildings and household locations they use wall attic spaces for their resting and roosting place

Management:

A thorough inspection is necessary to detect harborage sites. The odour and little spots of excreted blood assist in pinpointing these areas. Sanitation is helpful in control because it assists in increasing awareness of some harborage areas.

Bed bugs can be controlled with thorough applications of residual insecticides applied to cracks and crevices and into other known or suspected harborage areas. Furniture, especially mattresses, should also be lightly sprayed. However, the bedding of infants and sick/aged individuals should not be treated rather replaced. Vacuuming provides a major support in Bedbug management as they suck the eggs and young bed bugs hiding in cracks and crevices where the chemicals may not reach. Most situations one time application of pesticides may not yield the desired results in comparison to a two applications method with a gap of 15-20 days wherein the probability of the young ones hiding in cracks and crevices crawling out will again be susceptible to the chemical application.

8.

Black Carpenter Ant

Order: Hymnootera
Family: Formicidae
Scientific name: Camponotus spp

Description:

Carpenter Ants get their name from their nest building, where they excavate the wood and form smooth tunnels inside of the wood. They do not eat wood, only tunnel and chew through wood to create nests. The black Carpenter Ant indeed are the most thoroughly studied species in the West and it is now beginning to be addressed as a separate pest to be managed in India as well. Carpenter ants range from 1/8 to 1/2 an inch long. The Queens are slightly larger. Winged carpenter ants can be distinguished from termites by the larger size and shape of their antennae, waist and wings. The workers of an established colony vary in size; commonly, they are black; however, some species are red and black solid red or brown. They have one node in the petiole and a circle of tiny hairs on the tip of the abdomen. When seen from the side, the thorax is evenly rounded.

Biology:

The adult winged female or queen loses her wings soon after mating with the smaller males selects a secluded nesting site where she raises their first brood of workers. These workers are very small but assume the care of the larva and the queen following their maturation. Further workers are larger than those from the first brood, having received better care from the first brood of workers. All workers are wingless. Mature colonies range in size from several thousand workers, including satellite nests for some species an average of 10 to 20,000 workers and colony size goes up to 200,000 workers. Black Carpenter ants complete their lifecycle in about 60 days. Swarmer's do not appear in the colony for several years, usually three to four years or 6 to 10 years, depending on the species. Swarming for these ants occurs from May to August and End February to June, depending on the species.

Habitat and Economic Importance:

Carpenter ants are social insects that usually nest within the wood. They commonly excavate galleries or tunnels within rotting and sound trees and in structures readily infest would, foam insulation and cavities. They prefer to excavate wood damaged by fungus and often are found in conjunction with moisture problems. The workers excavate the nest, forage for wood and care for the young. Carpenter ants feed on sugar solutions from honeydew-producing plants. Carpenter ants feed on honeydew from aphids, sweets and the juice of insects they capture. They do not eat the wood as they excavate their nest. They actively feed at night well after sunset and continue feeding through the early morning hours. Foraging trails can extend up to 300 feet. Carpenter ants enter structures through gaps and cracks while foraging for food. However, the appearance of a large number of winged adults within a structure indicates that the nest exist indoors. The workers push wood shavings and pieces of foam insulation's out of the nest through slit like openings in the surface of the wood or other nesting site material. This material which may contain fragments of other insects and structural moisture problems, are clues to look for when trying to locate a colony within an infested structure. Rustling sounds in wall voids are another indication that there is a nearby colony.

Common Site of Occurrence:

Carpenter ant species reside both outdoors and indoors in moist, decaying, or hollow wood, most commonly in forest environments. Garden areas or greenery around the building or heavily tree surrounding structures are the most common areas preferred in outdoors. Indoors they prefer places such as under windows, roof eaves, decks and porches, chimneys sinks, door frames are more likely places to be found. They use tunneling system to live in trees which is also most common place of living. It is found in households, manufacturing units, educational institutions, homestays, resorts etc human inhabitation with greenery around.

Management:

Carpenter ant control can be very difficult and thus requires an integrated approach involving moisture elimination removal of overhanging tree limbs, stumps and firewood. Any mechanical alterations to prevent an entry is an important tool for the control of Carpenter ants. It is important to locate the source of the ants and their nest and satellite nest. Areas where water leaks occur, particularly the roof, bathrooms and kitchen, should be inspected. The most complete control is accomplished when the nests are treated with a residual spray or dust formulations. Control may involve drilling holes in hollow doors, Walls voids and ceiling voids exploited by Carpenter ants for nesting inside. In some situations, it can be useful to drill infested wood and apply a dust or liquid form directly into the galleries. Infestation can be reduced but not eliminated by treating the trails that the workers follow while foraging. Infestations that originate outdoors can be reduced by the application of barrier treatments. One can use Ant Gel Baits application which will not give a fast action result but will help to gain control over the colony, usage of the same can be evaluated based on the situation of management, time and adaptability of the customer.

9.

Black Widow Spider

Order: Aranea

Family: Theridiidae

Scientific Name: Latrodectus spp .

Description:

With almost 48,200 species of spiders spinning their web for millions of years, scientists have been curious and studying them across the world except the Antarctic. In the world of spiders, there exist many Spider species that are venomous too. Spiders found in India include 37 unique species from confirmed sightings by contributing members of Spider ID. It is important to remember that Spiders seen in India are not bound by the territorial lines decided on by humans. Therefore, their distribution is subject to change.

Red black or Black Widow (Latrodectus – Theridiidae family) is a native of Australia. There are records of this spider's appearance in most parts of Southeast Asia. In India, Black Widows are recorded in Gujarat state to a maximum extent with a good population. In fact, it is less documented in South India. After mating, the female spider feeds on male (Sexual cannibalism) hence, the name Black Widow Spider. Their venom is rarely fatal to humans, but its bite may cause complications depending on the immunity potential of an individual. Spiders intentionally never bite, but if we disturb them accidentally, they are likely to bite.

Biology:

Black Widow Spiders lay their eggs in silicone sacs which they protect in their nest. A female produces 6 to 21 sacs during her lifetime, each of which contains 200 - 400 eggs. The young spiderlings remain in the case until the second mold. They live in the vicinity of the nest for two to three weeks before producing long threads of silk which make them float away. Female spiderlings undergo 4 to 9 molts before maturing, a process that requires 54 to 107 days of development from egg to adult. Approximately one year female Spiders live up to three years and mail approximately 180 days.

Habitat and Economic Importance:

They prefer dry, dark locations to spin their webs. They often seek out warm areas during the winter months. Outdoors live in protected areas, i.e., stones and decks, firewood piles and hollow tree stumps. The preferred habitats are dry man-made structures, including barns, outhouses, henhouses, sheds, meter boxes, barrels and woodpiles. Indoors, they hide in sheltered, dimly lit locations and are often found in garages, basements and in crawl spaces. They prefer cluttered areas that offer more harborage for their prey. The females may bite in defence, especially after laying eggs. Symptoms of a black widow spider bite include fever, increased blood pressure, sweating and nausea. Pain is usually almost immediate and reaches its maximum in 1-3 hours. The pain associated with a bite may continue for one to three days and gradually subside.

Common Site of Occurrence:

They are generally and outdoor resting and breeding spiders, only some accidental or unforeseen situations only makes these spiders found inside the household. They prefer undisturbed and dry locations to catch their prey. Mostly found in among rocks and wood piles, under decks, hollow tree stumps, unattended basements, garages, pump houses etc. In the commercial establishments or a frequently functioning or activity based structures these spiders are commonly not seen.

Management:

Black Widow Spider control can be achieved by reducing the clutter in basements and garages, which, in turn, eliminates their hiding spots. When webs are visible, use caution before putting hands or feet in that area.

Outdoor lighting attracts insects to the structure and should be changed. Pesticide can be applied as residual sprays or dust formulations applied into potential or known harborage areas. Misting or ULV treatment with pyrethroids can be used to kill exposed spiders during spider webs and leave them undisturbed for several days also is a successful strategy.

10.

Bandicoots

Order/ family: Rodentia / Muridae

Scientific Name: Bandicota indica / Bandicota bengalensis

Description:

Bandicoot is normally rats found in the field or in burrows in the field. Bandicoot are also called as Greater Bandicoot rats for the species indica and they are also known as lesser bandicoot rats for species bengalensis. The greater bandicoot rat has dark grey-brown upper parts with a profusion of long, black hairs. Sides are grey with a few long, black hairs. Short, light grey fur occurs on the ventral surfaces. It has a dark and naked, scaly tail and dark feet with light-coloured claws. The young are much lighter in colour.

The Lesser Bandicoot rat (B.bengalensis) is a giant rat of south Asia and not related to marsipuals or true bandicoots. These are serious pests in cereal crops and the name Bandicoot is derived from the Telugu word "Pandikokku", meaning pig rat. They have a blackish to grey fur on the body and the uniform dark tail is shorter than the head-body length.

Biology:

Females of Bandicoot indica species lays 8 to 10 litres during their lifetime and 8 – 14 rats per litre. Young ones reach reproductive capability in 50 to 60 days from birth. The life span of adults is about a year. Females of B.bengalensis can have up to 10 litters and 10-12 young ones per litter. Young reach sexual maturity around 60 days after birth. The lifespan of adults is about 8–9 months. Young ones are born blind in both species. They feed on household refuse, vegetables, grains and are also a very serious pest problem in the poultry industry, damaging poultry feed and also poultry birds at times. Of all the three species, the lesser bandicoot is an especially aggressive burrower and has been reported to make tunnels in concrete cellars. They construct burrows to nest and bear their litter.

Habitat and Economic Importance:

Large Bandicoot are very aggressive in nature and if they are caged with other rats or species, they fight until death. They are also voracious eaters and they occupy the outskirts of human inhabitation like compounds, gardens, garbage etc., areas. They are so aggressive that, at times, even dogs and cats are afraid of these bandicoots. Their burrowing habit causes great damage to buildings and structures; they can also tunnel through brick and masonry. Commonly lives in cultivated plains and gardens and is one of the most destructive pests to crops and cultivation. It digs burrows with a characteristic pile of the earth around the entrance. The storage chambers in the burrows are stocked with large amounts of grain, especially during harvest time. The lesser bandicoot is nocturnal or most active at twilight. The burrowing nature of bandicoot at times is so heavy that it may cause damage or cracks in the structures / dwellings.

Common Site of Occurrence:

This is a species which is a completely outdoor living and need burrows to breed and rest. The common place of occurrence is the agricultural fields and open lands which is not inhabited. In the modern urbanization if there is a green zone which is encroached by a structure then the periphery of the structure and the vicinity of the structure can have the bandicoot populations.

Management:

Bandicoot being predominantly a field rat and also live in a burrow, an extensive rodent control program needs to be designed to control these rats. The trapping methods in these species infestation may not be that successful due to their aggressive nature and behaviour, which may cause them to escape and also, in this process, it may get hurt itself more. A complete thorough inspection of all the possible habitant areas inspection has to be done to ascertain the burrow population and thereby assess the extent of rat infestation. On an expert evaluation, even live burrow and the old burrow identification can also be made. Old burrows can be closed by mud/soil as in some cases these Bandicoot unused used burrows or in some cases used burrows are overtaken by filed snakes for their resting during egg-laying stages.

Exclusion is a critical aspect of rodent control. Since rats can squeeze through a half inch gap, anything larger must be sealed. Since they can chew holes, ceiling smaller holes should be considered. Sheet metal, cement, quarter inch hardware cloth, expandable forms comma etc., all the materials used of choice which can be used for service. Door sweeps should be installed around gaps on doors, windows, and other openings.

Several anticoagulant rodenticides are available as pirates, packets, and blocks. Baiting for Bandicoot gives better results as they consume a lot of bait due to their voracious eating habit. So, the baiting techniques should be done with a change of baits as much as possible and to use more of the locally available foods if baits are prepared with the concentrated rodenticide powder. The baiting program should continue for a week to 10 days to have better and complete control of the population. When baiting indoors, these products should be placed in tamper resistant bait stations and in areas that are then accessible to children and pets. Outdoors, palpates can be placed dipped in the burrow using a long-handled spoon and then the burrow should be closed. If packets and blocks are used in Burrows, there is a risk that the rat will push the product outside the burrow. Tracking powders are also available, which contain either an acute single dose toxicant or anticoagulant active ingredients and typically are used to dust Burrows and runs. Using gas in outdoor Burrows is an option but is risky, particularly if the borrower extends under or into a structure.

11.
Bottle Fly/ Blow Fly

Order/family: Diptera/Calliphoridae
Scientific name:
Phormia regina – Black Blow Fly
Phaenicia sericata – Green Bottle Fly
Calliphora – Blue Bottle Flies

Description:

Several different species of Blow Flies may infrequently infect structures. As a group, they are easily identified by the metallic blue, green or yellow-brown sheen of their stout bodies. They range in size from ¼ inch for the black blow fly to more than ½ inch for the blue bottle flies. These flies are active and usually seen buzzing at windows. The Black Blowfly is black with a bluish-green lustre (shiny); the green bottle flies are metallic blue-green; and the blue bottle flies are metallic blue. These flies have sponging mouthparts and one pair of wings. Mature larvae (maggots) are 3/8 – 7/8 inch long, legless, eyeless, and taper from the larger round rear segment to the pointed head, which consists of a pair of mouth hooks. They are cream-coloured and have large spiracles on the posterior plate.

Biology:

Depending on the species, females lay from 540 to 2373 eggs in their lifetime. Eggs are laid in batches of 100 to 180 on meat, fish, or carrion, but they are also attracted to animal manure, garbage, and rotting vegetable matter. The eggs hatch in less than a day. The larvae develop rapidly in this environment, molting three times in three to four days. The larvae typically leave the larval food to pupate in the soil for six to seven days. The larva and pupa are the overwintering stages. Development time (egg to adult) can be as short as 10 days but typically is 15 to 20 days.

Habitat and Economic Importance:

Blow flies seldom are a significant problem in structures but can be quite annoying because of their persistent buzzing. One of the first signs of and infestation is when the larvae leave the breeding medium to pupate. These flies frequently are found developing in the decaying bodies of rodents and other animals that have been killed or have died inside the attics, wall voids, or chimney of structures. Some species are strong fliers, and all are attracted to bright light. The major health concern associated with these flies is their ability to transmit diseases and cause human and animal myosis and infestation of living or dead tissue by fly larvae.

Common Site of Occurrence:

These insects are predominantly seen the vicinity of the dead animals or dead rodents. They are found in horse / cattle stables, garbage heaps / large garbage dumps, surrounding the structures. This is not seen as an insect inside the structures or facilities. Can be common flies in the large meat / fish market areas, meat or poultry processing centres, poultry farms etc facilities vicinity.

Management:

The initial inspection should focus on the identification of the fly (adult and/or larva) causing the problem and location of all resting and larval development sites. Adults often rest in breeding areas. Hence, it is helpful to inspect at night, as well. Sanitation, or source reduction, is the most important step in controlling them because it eliminates larval breeding sites. When successful, it significantly reduces the need for pesticide application. If trash receptacles are the problem, property owners should be instructed to empty and clean them at least weekly to disrupt the development cycle.

Mechanical control measures include insect-proof garbage containers, self-closing doors, screening, caulking and air curtains. Electric fly traps, sticky traps and other devices should be used to reduce adult fly populations indoors and out. Insecticide applications should be directed at adults because sanitation and removal are the best control measures for larval breeding sits. Baits, aerosols and residual insecticides are the most widely used pesticides. Ultra-low volume and aerosol applications should be made when the adults are most active and there is the least risk of drift, surface contamination, and human or animal exposure.

12.

Brown Dog Tick

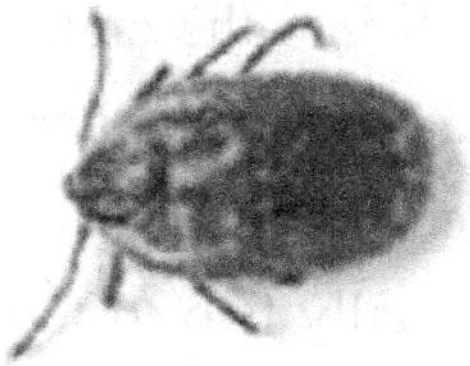

Order/ Family: acari/ Ixodidae
Scientific Name: Rhipicephalus sanguineus.

Description:

The brown dog tick is uniformly red-brown and 1/8 inch long. Its mouth parts are easily seen when viewed from above. The body is flat and shaped like a teardrop. The females' shield-like area remains unchanged, but the rest of the body stretches and changes from red-brown to blue-grey, the body size increases to a half inch long while feeding. The male has tiny bits scattered over the back. The adult has festoons (rectangular areas divided by grooves) along the posterior end of the abdomen.

Biology:

The female drops off the host dog, seeking a quiet location to deposit a mask of 1000 to 3000 eggs. She tends to move upward and often lay her eggs in cracks and crevices near the ceiling or roof of candles and, indoors, around wall hangings and the ceiling. The female dies soon after laying their eggs. In 19 to 60 days, the eggs hatch into six legged larvae, which then move down the wall and attach to a dog as soon as possible. Then, the larvae drop off and moult into the eight-legged nymphs, which in turn moult into the adult after another blood meal. The adults also feed on dogs. The entire life cycle can be completed in two months, and there are usually two to four generations per year. Adults can live unfed for 18 months.

Habitat and Economic Importance:

The brown dog tick is a parasite of dogs but is annoying and frightening to homeowners because often, it is seen on the walls and furnishings if the dog is infested. It seldom feeds on humans. It is a potential vector of Rocky Mountain spotted fever.

Common Site of Occurrence:

These insects needs body of the animal to complete their lifecycle hence they are found in very close vicinity of the dog or animals. It would be present on the body of dog for feeding on blood and later on it would be seen in cracks and crevices of the dog kennel, animal shed, voids and crevices inside the house if the dog is indoors. These insects is not seen in a commercial establishments except for veterinary clinics or hospitals, dog breeding and caring centres. Only if it is over populated it would be seen crawling inside the house holds or around the vicinity of the households.

Management:

Successful control of the Brown Dog Tick requires a three-step program that involves sanitation, tick control indoors and outside, tick control on the dogs. The infested house and/ all Kennel should be thoroughly cleaned in order to eliminate as many ticks as possible. Vacuuming is very effective indoors. Pet bedding and resting areas should be given special attention kennels, dog houses and structures occupied by the pets should be thoroughly treated to control ticks that have dropped off the dog and that reside in the harbourage areas. Residue sprays and dust should be applied carefully to all potential tick harbourage areas. It is important to remember that these ticks like to reside in the upper portion of the structures in cracks and crevices and areas frequented by dogs.

Outdoors, microencapsulated or vegetable powder formulation should be applied to grassy and bushy areas near the house or kennel, the edges of lawns and gardens, under porches, and other areas where the dog travels or spends time. Infected dogs should be treated by a veterinarian, a pet groomer or the owner on the same day on which the premises are treated pest control technician should never apply any product to an animal. Pets to be treated with Veterinarian recommended tick control medicinal bath.

13.

Brown Recluse Spider

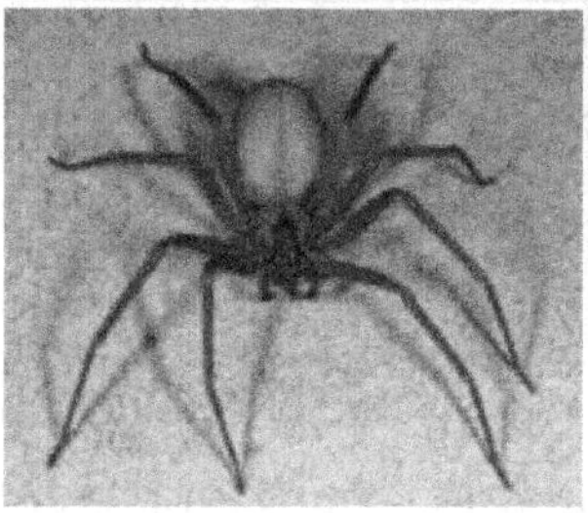

Order: Araneae

Family Loxoscelidae

Scientific Name: Loxosceles spp.

Description:

The Brown Recluse Spider, also known as the Violin Spider, is a venomous creature whose bite can cause children and adults to become ill. This insect is unusual because it has only six eyes (most spiders have eight) and wears a violin-shaped marking on its back. Six eyes are arranged in a semicircle on the top front of the first body segment. They have long legs that appear bare.

Biology:

About 40-50 eggs are contained within 1/3-inch diameter off-white silken egg sacs. The tiny emerged spiders gradually increase in size, molting five to eight times before becoming adults. The molted (shed) skins of the brown recluse have a distinct outstretched appearance and can be useful in confirming infestation. Brown Recluse Spiders mature in about a year and have an average lifespan of 2 to 4 years. The females produce up to 5 egg sacs in a lifetime. Infestation levels in homes vary greatly, ranging from one or a few spiders to several hundred.

Habitat and Economic Importance:

In nature, brown recluse spiders live outdoors under rocks, logs, woodpiles and debris. The spider is also well adapted to living indoors with humans. They are resilient enough to withstand winters in unheated basements and stifling summer temperatures in attics, persisting many months without food or water.

The brown recluse hunts at night, seeking insect prey, either alive or dead. It does not employ a web to capture food — suspended webs strung along walls, corners, ceilings, outdoor vegetation, and in other exposed areas are almost always associated with other types of spiders.

In homes, such webs are often produced by harmless cobwebs or cellar spiders. While sometimes considered a nuisance, spiders like the cobweb or cellar varieties prey upon other pests (including brown recluses) and, in this sense, could be considered beneficial. During daylight hours, Brown Recluse Spiders typically retreat to dark, secluded areas. They often line their daytime retreats with irregular webbing, which is used to form their egg sacs. Adult female recluses seldom venture far from their retreat, whereas males and older juveniles are more mobile and tend to travel farther.

Common Site of Occurrence:

Common spider occurrence in the household seen in the unused areas like news paper storage area, old clothes storage, attics, shelfs, godowns, and other areas in indoors. Outdoors it is more seen in the debris of stones, wood piles or unused materials dumped. In the commercial establishments like offices, educational institutions, hotels etc unused areas or left unattended can also have the occurrence of these spiders.

Management:

These insects are difficult to eradicate, largely because of their secretive habits. Virtually any dark, undisturbed area can serve as harborage, and many such places occur within buildings. Because of this (and the potential health threat), treatment is best performed by professionals. Likely hiding places include crevices, corners, and wall-floor junctures, especially behind clutter and stored items.

Reducing clutter affords fewer places for the spiders to hide and can enhance the effectiveness of treatments. They can also live behind walls and inhabit the voids within concrete block foundations. Outdoors, the spiders may be found in barns, sheds, woodpiles, and under anything laying on the ground. They also commonly reside behind shutters. Migration indoors can be reduced by moving firewood, building materials, and debris away from foundations.

Sealing cracks and holes in a building's exterior can further help to keep these and other pests outdoors. Some common entry points include gaps under doors, vents and utility penetrations, beneath the bottommost edge of siding, and where eaves and soffits meet the sides of buildings. Insecticides should be applied into cracks and other areas where spiders are likely to be hiding, attempting to contact directly as many as possible. Sprays of Pyrethroids may be employed. ULV Misting containing pyrethroids can be used to kill adult spiders.

14.

Brown Banded Cockroach

Order/Family: Blattodea/blattidae

Scientific Name:Supella longipalpa (Fabricius)

Description:

Brown banded cockroaches are half inch long when mature, light brown to brown, hand have two light, yellow-brown bands running across their bodies; hence, their common name. The pronotum has a dark brown area that is shaped like a liberty bell. Females are darker in colour and broader than meals; their wings cover only three quarters of their abdomens. The wings of the male completely cover their abdomens. The coloured bands are much easier to see on nymphs because they lack wings to obscure them. The purse shaped egg capsule is light brown, approximately a quarter inch long, slightly bowed, and typically has 7-9 eggs per side.

Biology:

The female brown banded cockroach produces about 14 egg capsules, each of which contains 14 - 18 eggs. The female carries take a capsule for 24 - 36 hours, then, using secretions from her mouth, attaches it to protected areas, the underside of the furniture, inside televisions and other appliances. Names emerge in about 70 days and molt six to eight times before becoming adults. This requires 90 to 276 days. The adults live about six months.

Habitat and Economic Importance:

Brown banded cockroaches prefer a warmer and drier environment than German cockroaches. Thus, they are not nearly so common in houses. They are found throughout the structures, preferring hiding places up off the floor, behind crown moulding, pictures, tapestries and other wall hangings, and in closets, furniture, appliances, computers, and telephones.

Common Site of Occurrence:

Brown-banded cockroaches are having higher preference for warmer and drier conditions compared to other cockroaches. They are found in indoors of households as well as any type of commercial establishments as well. They prefer hiding places higher of floors such as wall hangings, electrical conduits, switch boxes, electrical machines in the kitchens like grinders, ovens etc., computers, dining tables, picture frames, tapestries etc places.

Management:

Established populations of brown banded cockroaches are difficult to control because they are found throughout structures and not in areas commonly considered to be cockroach harbourages. Brown Banded Cockroaches prefer for feeding on the left over foods which if taken care can help in good extent to prevent these cockroach infestations. It is important to inspect thoroughly for these insects in order to find as many of their food sources and harbourage sites as possible. Cockroach monitoring traps are available to pinpoint sources of this cockroach infestation and to monitor areas about which occupants have a complaint, but infestations cannot be visually detected. Visual inspections can be conducted using a flashlight and flushing insecticides like propoxur can be used to flush cockroaches from their harbourages.

An effective cockroach management program depends on good sanitation to eliminate the food, water, and harbourage they need for survival. Clean-up to reduce cockroaches in the home and office environments must focus mainly on the residue in and around coffee machines, microwave ovens, stores, refrigerators, trash cans, furniture, and areas where exposed food is stored. In addition, it is critical to reduce clutter as cockroaches like to hide in stock boxes, cartoons and stored paper and cardboard materials, especially in dark, damp locations near food.

Vacuum cleaning can be used to physically remove exposed cockroaches. Infested equipment may have to be disassembled and may have to be treated for completer population removal before further use. The choice of treatment will depend on its potential effects on the equipment. Permanent reduction of cockroach populations can be achieved by sealing gaps/crevices to eliminate harbourage. To maximize effectiveness, paste, gel and dry baits should be applied with a syringe like dispensing that is close to harbourage sites. Large, plastic bait containers should be placed as close as possible to the dark concealed spots where cockroaches are living, preferably adjacent to edges and corners.

Large populations of cockroaches can be reduced or eliminated with the careful application of spray and dust in sensitive locations. However, these products cannot be applied inside electrical appliances or computers. Many insects break down rapidly, so an appropriate formulation for the environmental conditions must be selected.

15.

Cat / Dog Flea

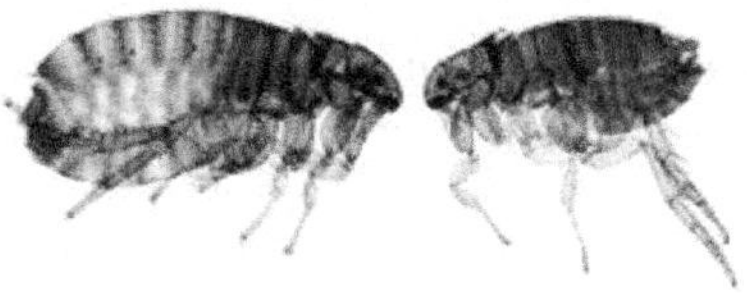

Order/ Family: Siphonaptera/ Pulicidae

Scientific Name: Ctenocephalides felis (Bouche) – Cat Flea

Ctenocephalides canis (Curtis) – Dog Flea

Description:

Cat and Dog Fleas can be found in the same area and are similar in appearance. They are small, 1/8 inch long, wingless, laterally flattened, having piercing-sucking mouthparts. The flea has very well-developed legs allowing it to jump at least six inches straight up. They are black to reddish-brown, and the body is covered with backward projecting spines, which help it to move between the hairs on the host animal. The head of the female Cat Flea is twice as long as it is high; the head of the female Dog flea is less than twice as long as it is high. Both Cat and Dog Fleas have a row of heavy spines on the front of the head (i.e., genal comb) and the back part of the first body segment (i.e., protonal comb). Cat and Dog Flea larvae are ¼ inch long when fully developed and look like Fly Maggots except for their well-developed heads. They have 13 body segments and are dirty white in colour with backward projecting hairs on each body segment. They have a pair of hook-like appendages on the last abdominal segment.

Biology:

Cat and Dog Fleas undergo complete metamorphosis. After each blood meal, females lay four to eight eggs at a time (but 400 to 800 total within her lifetime) on the host animal and/ or in its bedding. The eggs fall into the nest and/ or bedding of the host animal or wherever the animal happens to be at that time. The eggs hatch in about 10 days and the developing larvae feed on the adult flea faeces. Depending on the temperature, they molt three times in from seven days to several months. When mature, they spin silken cocoons in which they pupate. The pupal stage lasts up to 20 weeks. The adult Cat Flea often stays within the cocoon until vibrations stimulate it to emerge. Development (egg to adult) requires from 16 days to a year or more.

Habitat and Economic Importance:

Adult Fleas feed on blood with their piecing-sucking mouthparts and typically seek a blood meal within two days of becoming an adult. Cat and Dog Fleas prefer these two animals but readily feed on other animals, e.g., racoons, possums, rats, and humans. Adult Fleas remain on the host animal throughout their lifetime but are occasionally knocked off the animal by scratching. Occasionally, they can be found in the pet bedding and resting areas. Wild animals nesting in structures can initiate indoor flea populations. Larvae typically are found in areas where pets spend most of their time as well as in animal resting areas.

Common Site of Occurrence:

These insects are the ones which is normally on the body of the host like cats, dogs or other animals in their life cycle. Only on over populated or knocked off by the animal they are seen in the vicinity of the pets like bedding, cracks and crevices. These are not seen in commercial establishments. In households due to the over population of these insects it can spread to humans and can come in the vicinity of human resting or sleeping place for blood meal

Management:

Effective flea control requires customer cooperation and involves in three major steps: sanitation, insecticide application, and on-animal flea control. The house should be thoroughly vacuumed to remove larvae, pupae, and food materials. Vacuuming lifts the carpet fibres removing debris which, in turn, allows the pesticide to penetrate to the base of the carpet where the flea larvae are found. Pet bedding should be discarded or washed in hot, soapy water.

Indoor, residual insecticides in combination with an insect growth regulator which blocks the development of the flea larvae should be applied to carpeted areas, furniture where pets reside, and cracks and crevices on hard floors. Humans and pets should remain out of the treated area until all surfaces have dried and the area has to be ventilated. Outdoor areas which are frequented by the pet should be treated at the same time that the house is treated using microencapsulated or wettable powder formulations. These products should be applied either as a spot treatment to areas frequented by the animal or as a broadcast treatment.

The pet should be treated by a veterinarian, pet groomer, or the owner on the same day on which the house is treated. Numerous products are available for animal flea control, e.g., pills containing an insect growth regulator, spot on adulticides, flea collars, on-animal insect growth regulators, soaps, dips, etc. Regardless of the treatment, adult fleas must be eliminated from the animal in order for treatment to be effective.

16. Caterpillar

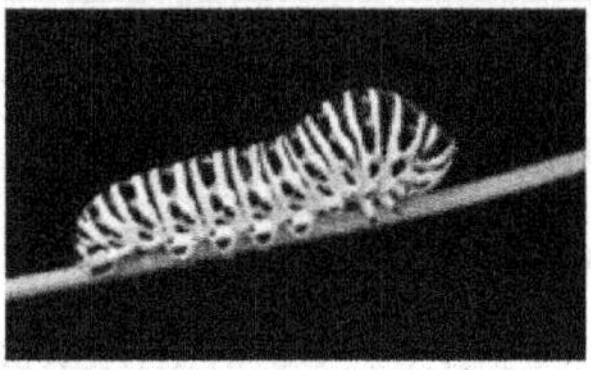

Order: Lepidoptera

Family: Various

Scientific Name : Various

Description:

Caterpillars are the larvae of butterflies and moths, which are members of the order Lepidoptera. Caterpillars may be tiny insect larvae, but they are voracious feeders, making them common plant pests. There are different types of caterpillars and vary in colour, size, diet.

Caterpillars have soft bodies and can be as small as 1 mm up to 14 cm. Like most insects, they have three body parts; head, thorax, and abdomen, an external covering called an exoskeleton. It has six pairs of small eyes known called stemmata which are arranged in a semicircle. They have tiny hair all over their bodies called setae and breathe through small holes called spiracles. They are able to sense touch through their setae and antennae. Their eyes are able to differentiate between light and dark but not detect images. Caterpillars have more than six legs, but most of those are false legs called prolegs. Prolegs help caterpillars cling and climb onto plants.

Biology:

Most species are herbivorous, meaning they feed on plants only. Most are considered destructive agricultural pests. Moth species, in particular, are known to cause extreme damage to agriculture and it is important for them to eat a lot, so they can sustain their growth into the next stage, which is adulthood. In their lifetime, they eat about 27,000 times their body weight. Many animals feed on caterpillars, as they are rich in protein. As a result, they have evolved various means of defence. Caterpillars have also evolved defence against physical conditions such as cold, hot or dry environmental conditions.

Common Site of Occurrence:

They mostly prefer to live on plants such as tree, grass, underneath leaves of shrubs etc. They usually need fresh and healthy food to survive. Due to the green around the structure or household, Moths or Butterflies if they have laid eggs inside the structure or on the structure then you would see the caterpillars developing on that. It would not come from somewhere but develop from eggs which were deposited there. Structures which are surrounded with lot of vegetation outside or any structure coming up in the midst of the jungle like vegetation will face this problem of caterpillar.

Management:

They can creep in through holes in screens, open windows or cracks in the walls. They chew on plants and can quickly put holes in indoor and outdoor foliage if the infestation is not treated. A broom can be used to brush caterpillars off the exterior and interior walls.

Synthetic pyrethroids can be used inside the house and for plants, a suitable crop-based insecticide that does not harm the plant has to be chosen. A quarterly preventive treatment will suffice for homes. For crops and plants, Bacillus thuringiensis is preferred by farmers as it does not harm birds, bees, fish, and mammals, making it an eco-friendly option for exterminating lepidopterans.

17.
Centipede

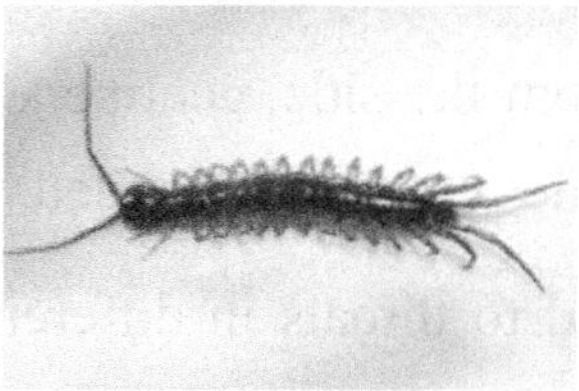

Class: Chilopoda
Family: Various
Scientific Name: Various

Description:

Centipedes are carnivorous and venomous. They sting and eat their prey, which typically consists of insects and worms. They are not aggressive towards humans but may bite if provoked. Centipede bites can be very painful, and the larger the centipede; the more painful their bite may be. Centipede bites rarely cause health complications in humans and are not typically dangerous or fatal. The bites can be identified by two puncture marks where they inject their venom into the skin. The area around the bite may become red and swollen.

Both Centipedes and Millipedes are made up of segments that link together to form one long body. With this body form in common, what are the difference between the both?

- Millipedes have two sets of legs per segment positioned directly under their body. Centipedes have one set of legs per segment positioned on the side of their body.
- Centipedes mostly eat insects after killing them with their venom. Millipedes feast on decomposing plants.
- If looking from the side, centipedes have a flatter body while millipedes are more rounded.
- They respond to threats in different ways. A millipede will coil up and release a smelly secretion. Centipedes can bite (which is typically harmless to humans) and run away quickly.

Biology:

Centipedes have a rounded or flattened head, bearing a pair of antennae at the forward margin. They have a pair of elongated mandibles and two pairs of maxillae. Many species lack eyes; some possess a variable number of ocelli, which are sometimes clustered together to form true compound eyes. Centipede reproduction does not involve copulation. Males deposit a spermatophore for the female to take up. The number of eggs laid ranges from about 10 to 50. The time of development of the embryo to hatching is highly variable and may take from one to a few months. Whereas under the right conditions, lithobiomorph species may reach a reproductive period in 1 year. In addition, Centipedes are relatively long-lived when compared to insects

(Lithobius forficatus may live for 5 to 6 years). Newly hatched Centipedes have 4 pairs of legs; during subsequent molts, they progressively increase the number of legs until becoming an adult.

Habitat and Economic Importance:

Female Centipede has been known to produce 35 eggs over a period of days. Centipedes are found in areas of high moisture, such as loose bark, rotting logs, under stones, trash or piles of leaves and grass. When they invade homes, centipedes are most commonly found in damp basements, crawlspaces, bathrooms or potted plants.

Centipedes are primarily carnivorous and obtain most of their moisture needs from their prey. Most house centipedes are nocturnal and prey primarily on flies, spiders and sometimes plant tissue, causing injury.

Common Site of Occurrence:

Centipedes prefer most environment and are active at nights. In indoors they are found in damp basements, bathrooms, cracks on the floor wall junction. In outdoors they are found under stones, decaying firewood, piles of leaves etc undisturbed places in the vegetation. This is not a common occurrence in the commercial establishments may be in any unattended vegetation areas on the periphery of the building.

Management:

Centipedes do not inhabit homes if it is bug-free. An effective method is to create an insect-free perimeter around the home.

Outside, leaf litter, grass clippings, firewood and building materials are to be moved away from the structure.

Centipedes need a high-humidity environment to survive. Hence it is recommended that the place is to be kept dry as much as possible. Sometimes, outdoor centipedes will come inside and can also crawl into the living area.

They can slip through extremely small holes, and it is recommended to use the pesticide free traps at night to catch these crawling ones. Pyrethroids spray and vacuuming can also give a good result. The perimeter spray should be done 3-to-10-foot band around the structure, into harborage areas and or around potential entry points.

18.

Cigarette Beetle

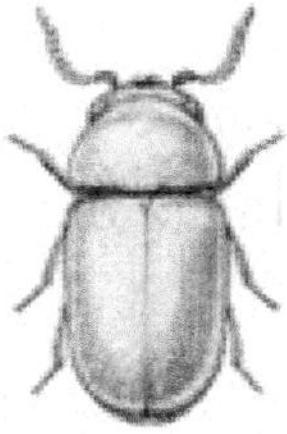

Order: Coleoptera

Family: Anobiidae

Scientific Name: Lasiodermaserricorne

Description:

The products infested by this pest include tobacco, cocoa, coffee, spices, dried vegetable substances, pulses, cereal products, pepper, ginger, turmeric, mustard seeds, oil seeds, oil cakes, raisins, dried fish, dates, drugs, garlic bulbs, books and furniture etc. It is a serious pest and can dangerously harm tobacco and cocoa. Small, robust, oval or nearly globular brick red to reddish-brown coloured beetle. 2.0 to 2.5 mm long. It is finely pubescent, they have a rounded, oval shape and the head is often concealed by the pronotum (plate covering the first region of the thorax) when the beetle is viewed from above. It has a humped appearance and the first body segment is bent downward and it has a saw-like antennae and smooth wing covers—mature larvae about 4.5 to 6 mm long, hairy with 3 pairs of thoracic legs. The first larval stage is much more active than the other stages. On hatching, the larva is translucent, gradually turning whitish and creamy, assuming a robust, wrinkled, grublike and curved.

Biology:

The adults are active fliers and do not feed. Their life span is short-lived and last in the range of 2 - 6 weeks. Females lay about 120 eggs over a period of 6 to 20 days, depending on the temperature. Eggs take 6 to 10 days to hatch at 30o C. The larval developmental period is significantly affected by the nature of the food supply at the completion of the larval period; an oval cocoon is formed in which the larva pupates. The pupal period is about 6 to 10 days. The beetle lies inactive within the pupal cell for 3 to 7 days; after emergence, it remains at rest for a day or two. The life cycle is 25 to 30 days, depending on the temperature and humidity available.

Habitat and Economic Importance:

It is believed to be a native of Egypt and has been recorded in the Pyramids, now is found world-wide. They prefer to reside in dark or dimly-lit cracks, nooks and crevices but become active and fly readily in bright, open areas, probably in an attempt to find refuge. They are most active at dusk and will continue activity through the night. Adults do not feed but will drink liquids. Cigarette beetles look almost identical to drugstore beetles but can be distinguished by two easily identifiable characters: the antennae of the cigarette beetle are serrated (like the teeth on a saw), while the antennae of the drugstore beetle are not and end in a 3-segmented club. They easily enter homes through open windows and doors and are frequently found along window sills. The presence of adults usually means the larvae are developing somewhere in the home.

Common Site of Occurrence:

It is a serious pest of tobacco and cocoa fields and storage areas. They are also found destroying pulses, cereal products, pepper, ginger, turmeric, mustard seeds, oil seeds, oil cakes, raisins, dried fish, dates, drugs, garlic bulbs, books and furniture etc. Warehouse and storage rooms or warehouse are the commercial places where these are found.

Management:

Preventing and controlling cigarette beetle infestations in the home is relatively simple; insecticides should be used only as a last resort. Locating the source of infestation is the first and most important step. Heavily infested items should be wrapped in heavy plastic, taken outside and thrown away. All food containers and items should be checked for infestation. Items can be placed in the refrigerator or freezer to kill all stages. To prevent reinfestation, clean up spilled flour, mixes, crumbs, etc. and thoroughly vacuum and clean areas where the contaminated items were stored. Store foods in airtight glass, metal or plastic containers.

Integrated pest management (IPM) programs are often implemented to control infestations at processing, distribution, and storage facilities. Residual insecticides registered for use on cigarette beetles can be applied to cracks, crevices and shelves in storage areas after the removal of stored products. Cigarette beetles are controlled in tobacco storage facilities with sanitation, supplemented by frequent application of insecticide and fumigation as and when necessary.

19.

Coffee Bean Beetle

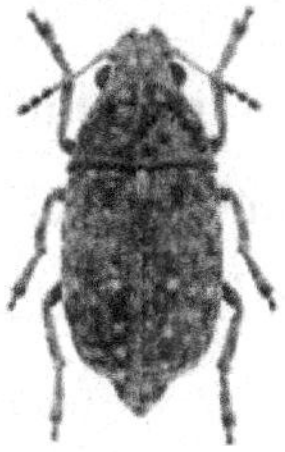

Order: Coleoptera

Family: Anthribidae

Scientific Name: Araecerus fasciculatus deGeer

Description:

The Coffee Bean Beetle or Coffee Borer Beetle (CBB) is an insect found around the world and prevalent in most coffee producing countries. It is among the most harmful pests in commercial coffee plantations and can attack 50-100% of berries on a farm if no control measures are applied. This type of beetle can feed solely on coffee beans. Other insects may occasionally nibble the seeds or other parts of the coffee plant but will need to eat other vegetation for sustenance.

The Beetles are usually tan, chestnut (dark) brown coloured, robust, oblong, 2.5 to 4.5 mm long beetles, the body is profusely clothed with setae. The prothorax is finely and densely punctate and the legs are large when compared to the body size. It breeds in all kinds of seeds, particularly coffee berries, cocoa seeds and corn, dried fruits and sweet potatoes, nutmeg, peanuts, spices, citrus fruits etc. It is a primary pest and a threat to the coffee crop.

Biology:

The CBB will usually enter the coffee cherry when the water content is 20% or higher and the cherries are still green. This is about 120-150 days after flowering and 30 – 150 days before harvesting. Matured larva 4.5 to 6.0 mm long, head of round, longer than broad, pale, straw -coloured, mandibles are stout and triangular, creamy white. Pupae are profusely clothed with setae, pupates inside or outside the seed.

Habitat and Economic Importance:

This is a strong flier and is frequently seen in the corn fields on the exposed and damaged ears of corn. It lays its eggs in the soft kernels of corn and coffee berries. The breeding continues after the harvest. When the corn or coffee berry becomes hard (< 8% moisture content), they are no longer attractive to the insect. The adults are attracted to coffee berries and beans or maize when their moisture content is between 12% and 20%. In storage, it can be bread on the maize flour, cocoa flour, tapioca flour etc. Oviposition - average 3 to 5 eggs per day, about 50 to 135 eggs per female, depending upon the moisture content of coffee beans. The eggs hatch between 5 to 7 days, larval stage lasts

46 to 66daysprepupal stage lasts 1 to 1.5 days and pupal stage 5-8 days. Both adult and larvae are destructive. Adult life span around 17weeks.

Common Site of Occurrence:

It is the primary pest and a threat for the coffee crop. It breeds in all kinds of seeds, particularly coffee berries, cocoa seeds and corn, dried fruits and sweet potatoes, nutmeg, peanuts, spices, citrus fruits etc. Warehouse and storage rooms or warehouse are the commercial places where these are found.

Management:

One of the most common ways to control the spreading of CBB is by doing preventive pickings towards the end of the harvest. This means picking is done in a way that no fruits are left in the trees or on the ground, regardless of their level of maturity. Insecticidal spray like pyrethroids is useful before females enter the berries. You will need a vacuuming done thoroughly in all areas of the foods stored, which would ensure all the eggs, larvae, adults and webbing are removed. Pheromone traps can be used to identify the area of activity if this is not clearly known. All infested items are to be discarded and un-infested products are to be kept in a secured insect-proof container, in the refrigerator, in the freezer. All cracks and crevices are to be treated thoroughly. Infestations in grain storage or grain processing facilities warehouses can be controlled using intensive sanitation and fumigation techniques.

20.

Common House Spider

Order: Araneae

Family Therdidae

Scientific Name: Achaearanea tepidariorum

Description:

The body of the female House Spider is about 3/16th to 5/16 inch long and yellow-brown with a dirty white abdomen, with stripes on the back. Males are smaller than females with a longer, narrower abdomen and feed on small insect's household pests such as flies, mosquitoes, ants and wasps. They can randomly attack grasshoppers, butterflies, cockroaches or other spiders depending on their size. Bigger females can also attract baby skinks inside their web by leaving fly remains hanging in it.

Biology:

House Spiders lay eggs in brownish silken sacks which have a tough papery cover. A female produces up to 17 sacks during her lifetime, each of which contains approximately 250 eggs. Young Spiderlings remain in the sack until the second molt; they are cannibalistic, eating one another. They live within the vicinity of the nest until after the second molt at which time they produce long threads of silk. Female spiderlings undergo 7 molts before maturing. Adults typically live for a year or more.

Habitat and Economic Importance:

House Spiders survive better in areas with high humidity, such as garages, sheds and warehouses. The lower humidity is not conducive to their survival. In structures with higher humidity, webs are constructed in upper corners under furniture around window and door frames, basements, garages and crawlspaces. Outdoors, webs are built around window and door frames near lights.

Common Site of Occurrence:

Common house spiders will live just about anywhere. They can be seen in gardens, backyards, basements, attics, barns, sheds, and any other type of man-made structure. These are the most common webs which we see in the urban inhabitation where onside of the web will be thick and the other side will be thin in size.

Management:

Inspection needs to be done to identify areas conducive to spider activity and recommended to remove clutter and debris, scrapped woodpiles, rocks and other protective outdoor materials from the structure. A vacuum cleaner should be used to remove spider webs, eggs and the bag immediately sealed and discarded. Using this technique to control house spiders is relatively easy since they build webs that are more exposed than those of black widows or brown recluse spiders.

Pesticides can be applied as residual sprays with special emphasis on application into potential or known harborage areas. ULV Misting containing pyrethroids can be used to kill adult spiders.

21.

Confused Flour Beetle & Rust Red Flour Beetle

Order: Coleoptera

Family: Tenebrionidae

Scientific Name : 1. Tribolium castaneum = Rust Red Flour Beetle

2.Triboliumconfusum= Confused Flour Beetle

Description:

These species are the commonest among the stored products insects in the world. They are found in flour mills, warehouses, grocery - stores, ships, etc. They are unable to feed on the undamaged, intact grains, and they depend on the primary pests. The adult beetles are 3 to 4 mm long, with elongate bodies reddish-brown in colour. Both species look similar but can be distinguished by the antennae. T.castaneum - compound eye, dorsal and ventral lobes separated by a comparatively broad notch. The dorsal lobe is not distinctly margined; the distance between the two ventral lobes is nearly equal to the transverse diameter of each ventral lobe. Antenna3-terminal segments enlarge abruptly. Terminal segment transversely round, wider than long. Mesosternum - Caudal median extension into basisternum absent. T. confusum – dorsal and ventral lobes separated by a narrow notch, Dorsal lobe strongly margined above. Distance between the two ventral lobes is two and half times as wide as the transverse diameter of each ventral lobe. Antenna segments gradually enlarge towards the apex, terminal segment is longitudinally round, with length width about the same.

T. castaneum: *Infects* Oil seed cake, groundnuts, cereals and, milled cereal products, dried fruits. **T. confusum:** Cereals and milled cereal products, dried fruits.

Biology:

Up to 300 to 450 eggs are laid by a single female. *T.castaneum develops more rapidly than T. confusum* and has higher fecundity. They limit their population by their cannibalistic behavior since both adults and larvae consume eggs and pupae. The eggs are laid at random in the produce (food material) over a period of many months and are oblong, 0.6 mm by 0.35 mm in size. The eggs are whitish, nearly translucent and sticky and hatch in 4 to 5 days. The larval period varies from 22 days at optimum conditions to over 100 days under adverse conditions. Pupation takes place in the produce; without the formation of the cocoon, the pupal period lasts 5 to 8 days and the emerged adults may live for as long as 18 months. Both adults and larvae feed on the produce. The Life Cycle for T.castaneumis30daysunder optimum conditions. In the non-preferred food like ground nuts, the life cycle at 25o C and 70% RH may take as long as 141 days, compared with 35 days on wheat bran. Larva for both species between 6 to 7 mm long, Elongate, cylindrical, dorsally convex, ventrally flattened, colour white-tinged with yellow. Body terga, dark banded. Head and terga of all segments clothed with yellowish setae. Pupa for both species Exarate, addictions type, elytral tip reaching the mid of 5th abdominal segment.

T. confusum cannot fly, a pest of flour mills in temperate countries and is very tolerant of dry conditions, develops successfully even at 10% RH. The development is possible between 20oC and 40oC. High humidity > 90% RH in combination with a temperature range between 30oC to 35oC are the most favorable for rapid development. Both the species **secrete defensive secretions** from their thoracic and abdominal glands, which are highly volatile. The secretions are toxic to its own species and possess bactericidal properties.

The adults are attracted to light but will go towards cover when disturbed. Typically, these beetles can be found not only inside infested grain products but in cracks and crevices where grain may have spilled. They are attracted to grain with high moisture content and can cause a grey tint to the grain they are infesting. The beetles give off a displeasing odour, and their presence encourages mold growth in grain. Reportedly only red flour beetles fly.

Common Site of Occurrence:

These species are the commonest among the stored products insects in the world and they are most common in tropic regions. They occur in flour mills, warehouses, grocery - stores, ships etc. They destroy Oil seed cake, groundnuts, cereals and milled cereal products, dried fruits, : Cereals and milled cereal products.

Management:

Flour beetles are one of the most common stored product pests found in houses. Infested flour or other stored products should be identified and discarded or heat-treated at 50 degrees Celsius for hours. The first step in managing an infestation is to find and remove the source of the infestation. Flour beetles can feed and survive on even the smallest bits of grain, so cleaning is a crucial part of controlling these pests. When attempting to locate the source, be sure to consider all likely food items. Once all of the infested material has been removed, be sure to vacuum and clean up the area around the infestation. If you have shelf paper, it would be wise to remove it, thoroughly clean under it with soap and hot water, and replace it with new paper. Storage areas and sites should be treated with Insecticide formulations as a supplement to the elimination of the source of infestations. Infestations in mills, food processing, storage and manufacturing facilities are controlled with sanitation, residual insecticide applications and necessary fumigation techniques.

22.

Cowpea Weevil & Bean Weevil

Order: Coleoptera

Family: Bruchidae

Scientific Name :

1. Callosobruchus chinensis = Cowpea Weevil

Callosobruchus maculatus = Bean Weevil

Callosobruchus analis = Bean Weevil

Description:

This common pest of stored legumes has a cosmopolitan distribution, occurring on every continent except Antarctica. The Beetle most likely originated in West Africa and moved around the globe with the trade of legumes and other crops. The ability to withstand a high degree of inbreeding has likely contributed to this species' prevalence as a pest. It is used as a model organism for both research and education due to its quick generation time, sexual dimorphism, and ease of maintenance. Products infested: Cowpeas, pigeon peas, green gram, dolichos beans, Bengal gram and some other pulses.

Biology:

All species are small beetles and relatively larger than the other stored product Beetle Pests, measuring 3 to 4.5 mm long. ***C.chinensis*** - Head free and deflexed; elytra elongate, oval, moderately shining. There is a white spot of setae on the post-pronotum, Scutellum is also white. Pygidium clothed with white pubescence. Males with deeply pectinate antennae; Females with serrated antennae. ***C.maculatus*** - Elytra in the female with dark black spots, in the males- the colour is overall brown with only small blackish markings. Elytra truncate do not completely cover the abdomen in both males and females; Pygidium in females bears dark coloured markings, whereas, in males, the markings are light coloured. Eyes large and emarginate; Antennae serrate in both the sexes. Larva for both species is Scarabaeiform and secondarily apodus. These are whitish, fleshy and thick. Head retracted into the prothorax. Pronotum with sclerotized teeth with single ocellus on each side of the head. Thoracic legs, when present, are like non-pigmented rudiments and are without claws.

Mature larvae pupate near the seed coat after making a circular, thinned out seed coat into a pupal cell keeping their head facing the 'window-door'. Primary pest of pulses. Larvae develop within the seed, causing serious damage.

Infestation may commence in the field before harvest. 70 to 100 eggs are laid attached to the pod or to the seed during storage. The eggs hatch between 4 to 5 days and the larvae which emerge into the seed directly develop up to the adult stage within the seed. Larval period 11 to 13 days; pupate near the seed coat in a pupal cell. The pupal stage of 6 to 8 days and infestation becomes visible in the form of 'windowed beans' (black windows). The adults are short-lived for about 9 to 12 days and do not feed on the produce. Development from egg to adult stage under optimum conditions takes 23 days.

Common Site of Occurrence:

These are majorly found in the fields and further the infestation may get carried to the storage areas like gowdowns, warehouses etc. They primarily prefer and found on pulses. Products damage are Cowpeas, pigeon peas, green gram, dolichols beans, Bengal gram and some other pulses.

Management:

Cleaning and sanitizing the area where dried food products are stored helps to deter Bean Weevils and other pantry pests. It is recommended to seal any cracks and holes in the home, and dried food products are to be stored in air-tight metal, glass or hard plastic containers. To provide control of exposed adults, residual insecticides can be applied to cabinets, closets and shelves where infested stored grains are found. All cracks and crevices are also to be treated along with any other potential harborage areas. Infestations in large storage facilities are successfully controlled with increased sanitation and bulk fumigation of the grain.

23.

Culex Mosquito

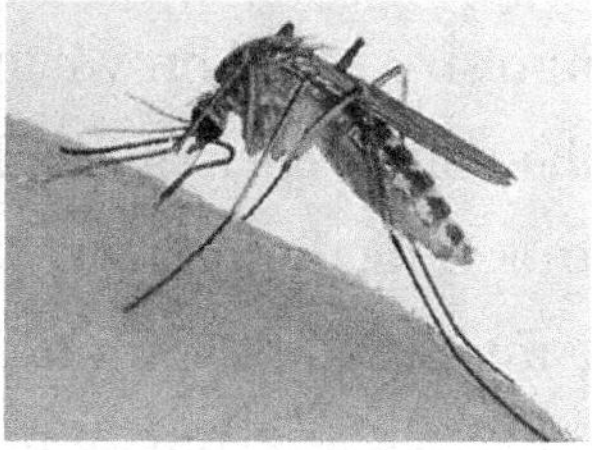

Order/ Family: Diptera/ Culicidae

Scientific name: Culex spp.

Description:

Mosquitoes are members of the order Diptera, i.e., flies. Mosquitoes are slender, long legged, two winged insects that are 1/8to ¼ inch long. They are unlike the other two flies as their wings and bodies are covered with scales, and they have long piercing and sucking mouthparts. *Culex* mosquitoes have blunt abdomens in contrast to the pointed abdomens of *Aedes* mosquitoes.

The larvae, are aquatic with well-developed heads, swollen and unsegmented thoraxes, and eight-segmented abdomens, which end in an elongated breathing tube. The pupae, or tumblers, are also aquatic and active. Their heads and thoraxes are encompassed in a large, oval mass with the slender abdomen attached.

The most notable species are Culex tritaniorhynchus, Culex vishnui, Culex pseudovishnui are the chief vectors of Japanes Encephalitis in India. The adults of these species are described below. They are medium sized, dull brown to brown-yellow, with even white bands at the end of each abdominal segment. The C. encephalitis mosquito is a medium sized dark mosquito with a white band across the middle of the proboscis (i.e., mouthparts). This characteristic separates it from almost any other Culex mosquito.

Biology:

Female southern and northern house mosquitoes lay their eggs in rafts which contain 50 to 200 eggs on the water surface. They hatch one or two days and complete development in from ten days to two weeks. Encephalitis mosquito adults overwinter in protected locations, and, after mating, lay their eggs in rafts that contain from 200 to 300 eggs on the surface of the water. The eggs, which are laid in all types of bodies of water, hatch in two to three days and, under ideal conditions, complete development within 18 days. These vectors are primarily outdoor resting in vegetation and other shaded places but in summer may also rest in indoors. They are in principally cattle feeders, though human and pig feeding are also recorded in some areas.

Habitat and Economic Importance:

They prefer to breed in water high organic matter content and that is contained in artificial containers such as cans, rain barrels, storm sewer basins, and in ponds and wells. The adults readily enter structures and are vicious biters who feed only at night. They rest within structures during the day. They feed extensively on cattle and are major pests in rural cattle sheds.

The encephalitis mosquito is a common pest and breeds in ponds, canals, borrow ditches, mud puddles, cess pools, barrels, bird bath, and catchment basins. These mosquitoes enter buildings in search of blood meal. Although they feed on humans and livestock, they prefer to feed on animals more. They are an important carrier of western equine encephalitis, a disease of the nervous system.

In the summer, mosquitoes become a major nuisance as they seek out humans for blood meals. Most service calls result when a customer plans activities in the backyard, such as an evening party, or there is a problem at the recreational sites. Perennial problems result from mosquito breeding sites located at inaccessible areas of the customer's premises such as salt marshes, swamps, storage sites, and in other difficult to manage sites, such as city sewers and under structures.

Common Site of Occurrence:

Culex mosquitoes breed in various types of stagnant water. Preferred habitats such as rainwater barrels, catch basins, storm drains, and septic tanks which are rich in organic material. Adults readily enter structure and vicious biters feed only at night. This is one of the prime mosquito nuisance found in the outside evening parties or recreation activities. At night they rest inside the structures only. They have breeding sites in inaccessible areas like salt marshes, swamps, city sewers, mid puddles, ditches etc places. Common mosquito nuisance in the Commercial establishments like offices, hotels, educational institutions, hospitals, eateries, etc.

Management:

It is important to identify the species which is causing the problems as well as the breeding sites. The most common methods of collecting mosquitoes for identification are the use of light traps, siphoning tubes at resting sites, dippers to collect larvae. Light traps can be small and portable or large and permanently wired. Both styles are designed to catch specimens without destroying them on sticky boards or by electrocution. Adults can be collected by using a siphoning jar trap. The best areas for collection during the date are in shady areas, such as inside and on the shady side of sheds and buildings, sewers, culverts, drainage pipes, etc.

Collection of larvae is more labour-intensive because it involves investigation of potential breeding sites which may not be accessible, as well as sampling the site in several locations using a dipper.

Lighting around the property can result in problems with mosquitoes. Because lights are beacons for mosquitoes, attraction to the area should be reduced by locating lives away from the structure and directing the mosquitoes towards them. Sodium vapour and/or yellow bug lights should be used instead of incandescent or mercury vapour lights.

Regardless of the species, the key to mosquito control is source reduction, i.e., the removal of the water breeding sites. For container breeders, this is usually resolved by site clean-up. However, flood water mosquito problems usually require the use of larvicide. The two most popular larvicides are the insect growth regulator (IGR) Methoprene and the bacteria Bti. Liquid, granules and extended-release formulations are available to treat the site even water is not present. When the product is exposed to water, the active ingredient is released into the water and kills larvae.

Adulticide with aerosol and ultra-low volume products is effective in killing the mosquitoes within the immediate area but does not provide long term control. Residual spraying of mosquito resting sites also reduces the problem. The use of repellents should be considered. Skin repellents, such as DEET and clothing repellents that contain permethrin, are very effective. At their best citronella candles and similar products provide temporary relief. Mosquito problems require considerable customer education because of the diversity of species, their often-remote breeding sites, and the ability of many species to travel long distances.

24.
Dampwood Termites

Order/ family:

Isoptera / Hodotermitidae/ Kalotermitidae/ Rhinotermitidae

Scientific Name:

Zootermopsis angusticollis (hagen)
– common Dampwood termite

Zootermopsis nevadensis (hagen)
– small or dark dampwood termite

Description:

Termites are social insects that live in large colonies. These are three castes: reproductives, workers and soldiers. Termite antennae have bead-like segments. The winged reproductive (swarmer) have a pair of long wings (equal in size) attached to the last two thoracic segments. The wings are broken off after swarming. The abdomen is broadly joined at the thorax, unlike the narrow abdominal attachment found on ants.

Damp wood termite winged reproductives are light brown with dark brown, leathery wings which can be up to one inch long. Some species have a fontanelle (i.e., a small opening on their heads) and/ or ocelli (i.e., simple yes). The nymphs, or workers, are 1/8 inch to 1/3 inch long and white to cream in colour with dark abdomens. The soldiers are 3/8 to ¾ inch long with large heads with long black-toothed mandibles. Their heads are black at the front and change gradually to red-brown, and their bodies are light brown. Faecal pellets are oval with flat sides, about 1/32 inch long and the colour of the infested wood.

Biology:

Most species swarm in the evening and are attracted to light. At most, damp wood termites produce a few hundred swarmers compared to subterranean termites, which produce thousands of swarmers. The founding male and female create a sealed chamber in the wood and, within two weeks, produce approximately 12 eggs. The second batch of eggs is laid the following spring. The colonies gradually increase in size, some reaching as many as 4000 individuals.

Habitat and Economic Importance:

Except for desert damp wood termites, damp wood termites do not require soil to wood contact, but the wood they infest must have a very high moisture content. These termites infest almost any kind of wood and tolerate very high levels of moisture. They often are found in dock and wharf pilings, logs, stumps and dead trees.

As they excavate wood while feeding, damp wood termites litter the inside of their tunnels with their faecal pellets. Damp wood termites usually enter structures in areas where there is wood soil contact and water damaged wood.

Common Site of Occurrence:

Dampwood termites normally do not require soil to wood contact. They are often found in the wood which have high moisture content. Often they are found in dock and wharf pilings, logs, stumps, dead trees, utility poles, fences etc wood structures. Indoors they are found in crawl spaces, attics and basements where there is water damaged wood.

Management:

Infested structures should be inspected for wood-soil contact, areas where there is a constant water supply, such as plumbing leak, damaged wood, and faecal pellets. The most effective methods of controlling damp wood termites are eliminating the moisture from the wood and removing infested wood and replacing it with treated lumber. Wood remaining in place and galleries can be treated with residual insecticides such as disodium octoborate tetrahydrate (Boron Powder), which is particularly effective when the wood has high moisture content.

25.

Drain / Moth Flies

Order / Family: Diptera/ Psychodidae

Scientific name: Psychoda spp. And Telmatoscopus spp.

Description:

Moth flies are small, delicate and "hairy", about 1/16 to ¼ inch long. They are yellowish, brownish-grey, or blackish; their pointed leaf-shaped, white speckled wings are held roof-like over their backs when at rest. The body and wings are covered with long, dense hair. These flies have long antennae with 13 to 15 bead-like segments. Mature larvae are 1/8 to 3/8 inch long. They are aquatic, long and cylindrical, with the fore end of the body somewhat flattened on the lower side, which has eight suckers. They have dark, hardened patches on the back of each segment, and they breathe through a hardened, stalk-free siphon tube at the end of the body.

Biology:

The females lay 30 - 100 eggs in the jelly like a film that covers the stones in the sewage treatment plant trickling filters or that line the water-free portions of drain pipes. The eggs hatch in two days and the larvae complete development in nine to 15 days. The larvae feed on algae, fungi, bacteria, and sludge. The pupal stage lasts about a day and a half. The development time (egg to adult) is seven to 28 days. Adults live up to two weeks.

Habitat and Economic Importance:

Moth flies or Drain Flies become an annoyance within some structures when they breed in the liquids found in drains, dirty garbage containers, and septic tanks. They breed in large numbers in filter beds of sewage treatment plants. Adult flies are found in great numbers on walls are active at night, seen hovering near the breeding site. During the day, they rest on vertical surfaces indoors and in protected areas outside. Adults feed on nectar and polluted water.

Common Site of Occurrence:

These flies are a common occurrence in commercial establishments in toilets, kitchens where there is clogged drains which has filthy clumps of organic waste with liquid. They breed in this organic slump which is seen both in commercial establishments as well as in households where you have such clogged drain conditions.

Management:

Infestation which develops in drains can be eliminated by scrubbing the area with a brush and sink-cleaning materials followed by hot water and a suitable drain de-clogging chemical treatment. It is important to scrub the drain all the way to the trap. If the trap is not holding water, it should be repaired. To prevent water evaporation in the trap of seldom-used drains, a small amount of vegetable or mineral oil should be placed on the surface of the trap water. Large numbers of adult flies are easily controlled using aerosol or spray applications of any pyrethroid group chemicals. Large populations of flies that are developing in sewage filter beds can be controlled with periodic flooding of the bed, which eliminates some of the larvae. Weed control in this area reduces the availability of resting sites. To control adult flies at outdoor resting sites, treat the area with a wettable powder or microencapsulated insecticide.

26.

Drywood Termite

Order/ Family: Isoptera / kalotermitidae

Scientific Name:

Incisitermes minor (Hagen) – Western Drywood Termite

Incisitermes snyderi (Light) – Southeastern Drywood Termite

Description:

Termites are social insects that live in large colonies. There are three castes: reproductive, workers and soldiers. Termite antennae have bead-like segments. The winged reproductive (swarmer) have a pair of long wings (equal in size) attached to the last two thoracic segments. The wings are broken off after swarming. The abdomen is broadly joined at the thorax, unlike the narrow abdominal attachment found on ants. Drywood

Termite Swarmers are light yellow-brown and the wings have three or more dark veins on the leading edge. They are 3/8 to 5/8 inch long and have red-brown heads and thoraxes. The nymphs are creamy-white with a yellow-brown head. The soldiers have large, parallel-sided, red-brown heads with massive mandibles which have an unequal number of teeth, and their bodies are light coloured. Faecal pellets are hard, elongate-oval shape, 1/25 inch long, and have blunt ends and size concave sides.

Biology:

Drywood Termites are a non-subterranean species, i.e., they neither live in the grown nor maintain contact with the soil; they do not build mud tubes. The male and female swarmer locate a crack or knothole in a suitable piece of wood; gnaw a small tunnel; seal the entry; excavate a small gallery; and mate. The first year, they produce a few eggs, one soldier, and 20 nymphs. In subsequent years, the colony slowly builds until after 15 years; a mature colony could have 3000 individuals, much smaller than subterranean termite colonies. There is no true worker caste; the nymphs perform their tasks for the colony. After seven molts, the nymphs become either adult reproductive or soldiers.

Habitat and Economic Importance:

Drywood Termites are found primarily in the southern United States, including the Southwest. They are occasionally found in northern states when the wood (e.g., furniture) they are infesting is moved into these areas. Colonies do not survive unheated structures. Drywood termites can establish a colony in totally non-decayed wood and maintain residence as long as the wood lasts; colonies are found in various locations, e.g., attics, door and window frames, trim, eaves and furniture.

Drywood termites feed across the grain of the wood, creating chambers and/ or galleries that are very clean and appear to be sanded smooth. The first signs of infestation are swarmers or the accumulation of faecal pellets below "kick out" holes.

Common Site of Occurrence:

Dry wood termites are usually seen in the wood that is mostly used for the structural furniture's or frames or any other usage. The entry will not be from the soil but either the elates or swarmer's settle their colony in non decayed wood. These termites live in above ground wood structures. Indoors it is found majorly in wood siding, wood roofs, framing, attics molding and window framing, joints and crevices.

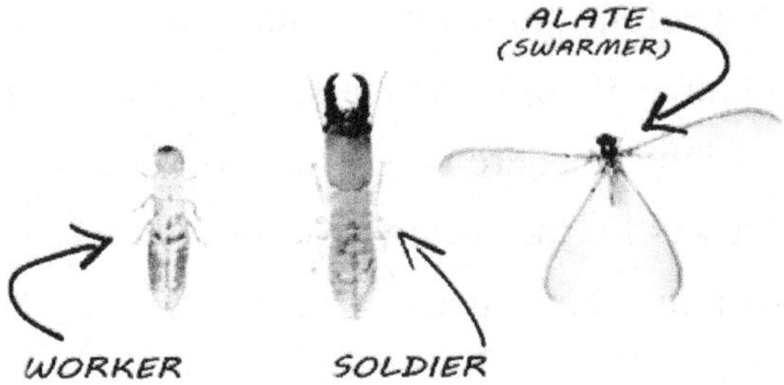

Management:

During the inspection, signs of infestation, particularly in attics, should be identified. Occasionally, infested wood can be removed and replaced. If the reinfestation is a concern, soffit and ridge vents should be screened in order to prevent re-entry of swarmers and exposed wood should be treated with disodium octaborate tetrahydrate (Boron Powder), Drywood termites are most effectively controlled using fumigant, either sulfuryl fluoride or methyl bromide. This involves tenting the structure and maintaining the concentration of the fumigant or a predetermined amount of time. Heat treatment is a relatively new strategy that requires tenting the structure and achieving and maintaining an interior wood temperature of 50° C for 30 minutes. Some items maybe damaged by higher core temperatures in the structure.

A number of spot treatments can also be done with EC based Residual insecticides available in the market depending on how much localized is the infestation and the quantum of the infested material. The judgment of the residual insecticide spray or the fumigation can be decided upon inspection and assessment of the quantum of infestation.

27. Dust Mites

Order/ Family: Acari/ Pyroglyphidae

Scientific Name: Dermatophagoides.

Description:

House dust mites are very small creatures about 1/64th inch long. Their presence is often suspected before they are actually seen and identified. The two most common species are *Dermatophilosis's* farina, the North American house dust mite, and *D. pteronyssinus,* the European house dust mite. Both of these are found worldwide. Adult mites are clear, and their cuticle has simple striations that can be seen from both the top view and bottom view. The bottom view of the mite also reveals long hair extending from the outer margins of the body and shorter hair on the rest of the body.

Biology:

The adult female lays 66 - 68 eggs singly or in small groups, after which when the eggs hatch, a six-legged larva emerges. After the first moult, and eight-legged nymph appears. After two nymphal stages occur, 8-legged adult appears. The life cycle from egg to adult is about one month. The adult lives for an additional one to three months. Their primary food source is dander from humans and animals, but pet food, fungi, serious, food stains and crumbs also provide needed nutrients.

Habitat and Economic Importance:

House dust mites inhabit buildings and micro-habitats that sustain a relatively normal to high moisture content. Occupied rooms that are dry and well-ventilated probably would not support this mite, whereas a damp room will support a mite population. The mites, their body parts and excretions are suspected of being allergens to over 500 million people worldwide, including 50- 80% of asthmatics.

Common Site of Occurrence:

Dust mites are crawling insects which needs a safe and secured place to shelter which are Carpeting, upholstered furniture, drapes, curtains, stuffed toys, and other fabric-covered furnishings. Carpet is a perfect breeding ground for dust mites. A favourite food is dander (both human and animal skin flakes) which is available in the carpets and mattresses.

Management:

The house dust mite is virtually invisible to the naked eye. Inspection for identification of these tiny arthropods and their body fragments requires a microscope to search through dust debris collected on cushions and mattresses, in pet bedding, cracks and crevices, and other areas where dander can collect. Mite population can be eliminated by reducing the humidity below 45% and improving ventilation. However, the allergenic materials persist even if the population is destroyed. Allergy reduction is dependent on removing these particles. It is extremely difficult to remove every packet of dust or to stop its daily accumulation.

Unless equipped with a HEPA filter, vacuum cleaners blow microscopic mite fragments into the air, often aggravating the allergy. Nevertheless, effective dust control measures do exist. Some furnace and air conditioner filters trap dust with an electrostatic charge. An allergist may recommend that customers reduce dust accumulation by replacing carpeting with wood, tile, or linoleum flooring. Currently, a few pesticides are labelled for house dust mites. Several non-pesticide products are available for the treatment of house dust mites and their allogeneic components; however, their effectiveness is variable. Hence, preventive measures and non-chemical methods proves to be more effective and long-lasting.

28.

Earwig

Order: Dermaptera

Family: Various

Scientific Name: Various

Description:

The Earwig gets its skin-crawling name from long-standing myths claiming the insect can climb inside a person's ear and either live there or feed on their brain. The common Earwig is an omnivore, eating plants ripe fruit as well as actively hunting arthropods. To a large extent, this species is also a scavenger, feeding on decaying plant and animal matter if given a chance.

Adult Earwig is a quarter to an inch long dark brown to black, the head being red and legs yellowish-brown. They are normally flattened in body structure. Earwigs have two pairs of wings, thread-like long antennas half the size of the body length. The characteristic feature of Earwig is the forceps like structure on the abdomen, which is used to capture the prey, defend themselves and fold their wings. The forewings are short oblong leathery plates used to cover the hindwings like the elytra of a beetle rather than to fly. The hindwing is a very thin membrane that expands like a fan, radiating from one point folded under the forewing. Even though most earwigs have wings and are capable of flight, they are rarely seen in flight.

Biology:

Earwigs undergo an incomplete metamorphosis developing through a series of 4 to 6 molts. The developing stage is called as instars. The female lays several batches of approximately 50 eggs in a nest like shallow depression beneath a stone. The eggs hatch in about 7 days. The mother may assist the nymphs in hatching. When the nymphs hatch, they eat the egg casing and continue to live with the mother. The nymphs look similar to their parents, only smaller after five to six instars; the nymphs will molt into adults. The male's forceps will become curved, while the females' forceps remain straight.

Habitat and Economic Importance:

Most Earwigs are nocturnal and inhabit small crevices, living in small amounts of debris in various forms such as bark and fallen logs. In indoors usually found in cracks and crevices, under furniture, underneath the carpet. They emit foul odour in their presence indoors. Some species are attracted to light.

During summer, they can be found around damp areas such as sinks and bathrooms. They tend to hide in shady cracks or openings or anywhere that they can remain hidden during daylight. There is a debate whether earwigs are harmful or beneficial to crops, as they eat both the foliage and the insects eating such foliage, such as aphids, though it would take a large population to do considerable damage. There is no evidence that they transmit diseases to humans or other animals. Earwigs have been rarely known to crawl into the ears of humans, but they do not lay eggs inside the human body or human brain as is often perceived.

Common Site of Occurrence:

Earwigs are mostly found outdoors and rarely its occurrence happens indoors. This may be found in protected, moist environments in leaf litter and all kinds of debris on the ground, under bark, under stones, between leaves and crevices. Earwigs are usually nocturnal and attracted to lights at night. It can get into cushions or beddings which are left outside overnight.

Management:

Earwig control mainly starts from outside than indoors. Remove all the moist harborage areas like wood piles, landscape timbers, decorative stones, rocks etc. The greenery should be properly mowed and trim the overhanging branches. Examine gutters and downspouts to make sure they drain away from the foundation. Adjust outdoor lights to shine from the yard onto the house – insects will be attracted away from the house. If moving outside light fixtures is not practical, consider changing light bulbs to yellow bulbs since white lights are more attractive to insects.

Entry points to the buildings should be sealed with a caulking of gaps around siding, windows, doors, pipes, wires etc. Pesticide treatment outdoors, which has a long residual action like Wettable powders, Encapsulated formulations are more effective for Earwig control. The product should be applied as a band of 3 – 10 feeds depending on the area from the perimeter of the structure into harborage sites and potential entry points.

29.
Field Cricket

Order: Orthopaotena
Family: Gryllidae
Scientific Name: Gryllus spp.

Description:

Field Crickets are large (15 to 31 mm), dark, and usually found on the ground. Male crickets use species-characteristic calling songs to attract sexually responsive females. Field Crickets are usually black but can also be brown or straw coloured. They have slender antennae which are longer than the body. They have large hind legs (for jumping) and most have well-developed wings. Nymphs are similar but are smaller and lack wings. Nymphs look like an adult but are smaller and the wings are not fully developed.

Biology:

The eggs are usually laid in the soil and the newly hatched nymphs burrow to the surface. They molt 8 to 10 times over a period of 2 - 3 months before becoming adults. Usually, the egg hibernates in the winters to favourable conditions. Each female lays 150 - 400 eggs and there may be several generations in a year.

Habitat and Economic Importance:

Field Cricket can be an important insect in agricultural and household problems, especially in summer when they move from fields to buildings. They can damage fabrics, especially if soiled and may chew on wood, plastic, rubber, or leather goods. They do not survive for a long time indoors and are active in the night, hiding in dark warm places during the day. They are attracted to light in large numbers.

Common Site of Occurrence:

As the name indicates it is most commonly found in the greenery outside lawns or such areas. Only during conditions when they enter the structures in search of moisture. Crickets are found on the soil, hiding under dead plants or on live plants. They only occur where there is plant material to eat, and they are most diverse and abundant in humid areas with lots of plants. They are active in night and hide in dark warm places in day time and are attracted to light.

Management:

The elimination of piles of bricks, stones, wood, or other debris around the home will help reduce numbers. Weeds and dense vegetation around the foundations of homes are other good hiding places. Nearby trash dumps, which provide both food and shelter, should be cleaned out. Since crickets are attracted to lights, the elimination of light sources at night will reduce the number attracted to the home area.

Measures such as caulking, weather stripping, and making sure all screens and doors are tight-fitting will help reduce the numbers that can enter your home or business. Adult Crickets can be difficult to control. Inside homes or buildings, ready-to-use sprays or aerosols applied to baseboards, door thresholds, and cracks and crevices where crickets hide will normally control them. Insecticide formulations like Wettable Powder or Micro-encapsulated products are more effective in the moist habitats.

30.

Fire Ants / Red Ants

Order: Hymenoptera

Family: Formicidae

Scientific Name : Solenopsis spp

Description:

Not all Red Ant species are Fire Ants, and not all Fire Ants are red in colour. In fact, most ants are not actually red. The ant often called the "red ant" is the **Red Imported Fire Ant.** Fire Ants sport **a red or reddish-brown appearance** and belong to the genus Solenopsis. Red Imported Fire Ant workers range in size from small to medium, making them polymorphic. Workers measure between 2.4 and 6.0 mm. The thorax lacks spines and the pedicle has two nodes. These colonies can contain several hundred thousand ants, including at least one queen.

Biology:

Fire Ant nests contain as many as 50,000 - 200,000 ants. There might be one or more queens in the colony. Female workers maintain the nest's structure and feed their young. Male drones breed with the queen or queens. When young queens mature in communities with more than one queen, they fly off with males to create new nests.

Researchers found that it would take two weeks of temperatures below -12°C to kill an entire colony. While they kill and eat other insects like regular ants, they have also been known to live on crops and animals. Fire ants can even form nests on water and float them to dry locations. Queens can produce approximately 1500 eggs; the larva matures into workers in about 25 to 40 days. Minor workers live 30 to 60 days, intermediate workers live 60 to 90 days and major workers live up to 180 days. Queens live for two to six years; males die shortly after swarming; in a year, there can be 6 to 8 swarms. Fire ant larvae are fed by the worker adults and eat only a liquid diet until their third larval instar is complete. Fourth, instar larvae are capable of digesting solid foods.

Habitat and Economic Importance:

They sting mostly on the legs racing upwards when their nests are disturbed. Each one can sting several times. The identification of fire ant stings is swollen red spots that develop a blister on the top. Stings hurt, itch, and last up to a week. Fire ants usually list within the ground but can develop colonies within structures, especially in the areas near the soil.

They are attracted to electrical junction boxes; when nesting within the soil, they build large unsightly mounds, which are a determinant to the cultivation of fields. They also gnaw on electrical wiring and clothing, especially if the clothing is soiled. They prefer high protein foods but will feed on practically everything, including other insects. Fire ants are omnivorous, and known to eat meats, greasy and sweet materials, feed on animal or vegetable sources of food.

Common Site of Occurrence:

These ants usually nest within the ground but can develop colony inside the structures in areas near soil. Found and attracted to electrical junction box, air conditioners and in soiled clothes unattended. They are predominantly found outside in soil mounds as they feed on honey dew, dead insects, seeds, fruit juices etc high protein feed available outside.

Management:

Applying Barrier residual insecticide spraying of any synthetic Pyrethroid, especially in areas where ants appear to be entering or travelling, can be a controlling mechanism. Quick control is also obtained from Propoxur Insecticide treatment in the travelling and nest building areas of ants. If proper identification of the ant route and the colony is not made and sporadic insecticide application could cause in spreading of the population to different places.

31.

Flat Grain Beetles

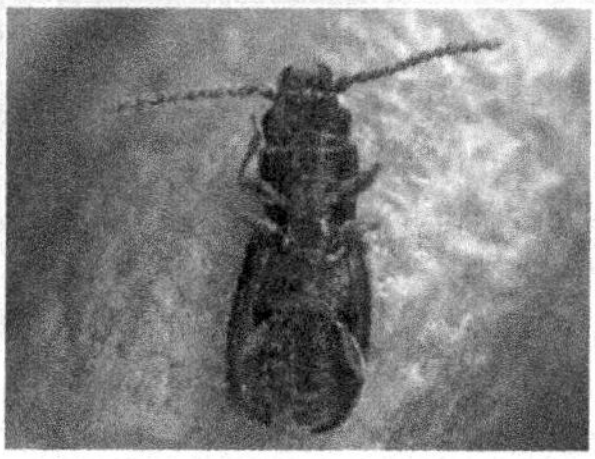

Order: Coleoptera

Scientific Name: Cryptolestes pusillus

Description :

Adults are small (1.5 to 2.0 mm), reddish-brown and flattened with long bead-like antennae. Adults have a ridge on the head and prothorax. Larvae are elongated and flattened with well-developed legs. Larvae are pale, white to yellow in colour, with a darkened head capsule. Elytra broadly rounded at the tip almost truncate covering the abdomen, emarginate, striate, punctate, slightly longer than the head and thorax together. The pupa is Exarate type that pupates in a sticky cocoon with adhered food particles. Products infested are cereals, oil seeds, milled cereal products, cocoa, cowpeas, copra etc. Primary pests of dry whole grain, also of damaged grain and milled products.

Biology:

These small beetles are scavengers and are capable of breeding in produce that is dusty, contains broken grains and has a high moisture content or is already infested with insects. The larvae show a marked affinity for the wheat germ. Females lay nearly 200 eggs in the grain commodity, often in cracks or split in the grains and hatch into larvae which are long, slender and straw coloured. When fully grown, they spin cocoons of a somewhat sticky substance to which particles of food adhere. Both adults and larvae are destructive, but the growth of molds in the endosperm of the cereal converts the food material into a more suitable larval food. Life cycle 23 days at the optimum of 33o C. In this case, humidity plays an important role in the survival of the species. Mortality is high when RH is less than 50% and below 25% development ceases altogether. This takes 21 days to reach the adult stage at 37o C. Adults are long lived, nearly 6-9months.

Common Site of Occurrence:

These small beetles are scavengers and are capable of breeding in produce which is dusty, containing broken grains and has a high moisture content or is already infested with insects. The larvae show a marked affinity for the wheat germ. These destroy Cereals, oil seeds, milled cereal products, cocoa, cowpeas, copra etc. Primary pests of dry whole grain, also of importance as pests of damaged grain and milled products.

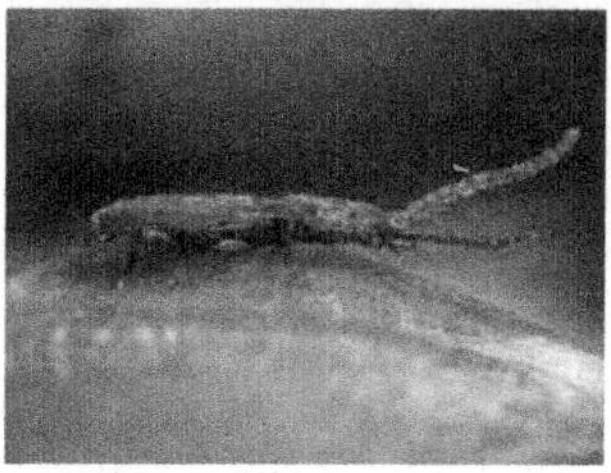

Management:

Damage is not readily identifiable as being directly caused by this species. Damage is usually restricted to grain stored under high moisture conditions. Both larvae and adults feed on germ and endosperm. Heavy infestations may cause the grain to heat and spoil. Damage may spread fungal spores through the grain. To control Grain Beetle infestations, you will need to empty out your entry pantry and discard all of your stored products and conduct a thorough vacuuming. After that, wipe all the shelves, apply insecticide to destroy adult Grain Beetles and halt the development of eggs and larvae.

32.

Fruit Fly

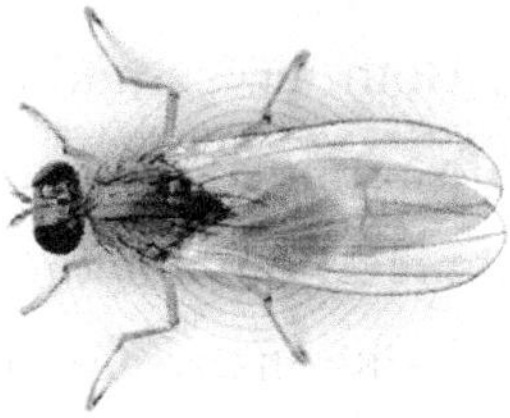

Order/ family: Diptera / Drosophilidae

Scientific Name: Drosophila spp.

Description:

Adult Fruit Flies are 1/8 inch long, dull yellow-brown to dark brown. Some species have distinctive red eyes; the wings have two "breaks" in the leading edge (vein) near the body. The larvae are small (1/10 to 1/5 inch long) and very distinctive with an extended, stalk like breathing tube at the rear of the body. The pupae are brown with two horn-like stalks at one end.

Biology:

The eggs are laid onto the surface of fermenting fruit or vegetables or areas where moisture and yeast are abundant. Each female produces up to 500 eggs, and they hatch within 30 hrs. The larvae complete development in five to six days and crawl to drier areas of food or elsewhere in order to pupate. The life cycle (adult to adult) requires eight to ten days.

Habitat and Economic Importance:

Fruit Flies are common structural pests frequently associated with fermenting fruits and vegetables. They easily develop in over-ripe fruits or other foods, fermenting liquid in the bottom of garbage cans, a dirty mop, or a rotting potato or onion in the vegetable bin. Recycling bins fruit and salad bars are ideal habitats and have resulted in increased problems with this pest fly.

Common Site of Occurrence:

Fruit flies are common pest of indoor structural pest and are very rarely seen outside. The breeding and presence of these insects are closely associated with fermenting fruits and vegetables. They easily develop on over ripe fruits and foods, fermenting liquid and garbage cans, dirty mops, rotting potato or onion in the vegetables. These insects are not found in clean and hygienic places. It is found in both commercial establishment's food handling areas and households.

Management:

They are best controlled by finding and eliminating the breeding material. The presence of adult flies indicates that larvae are developing in nearby fermenting materials. Complete and thorough sanitation is necessary to eliminate the source of infestation. Insect light traps and baited jar traps fitted with tops permit fly enter and prevent escape are effective in reducing the population. Several species are small enough to pass through typical screens, so a smaller mesh may be required. In limited situations, an aerosol can be used to knock down adults. Any spraying activity, care has to be taken to ensure proper protection of food material as they hover around the food material in very close proximity. The most successful and effective control methods have always been, in this case, is, source reduction methods coupled with good sanitation methods.

33.

Fungus Gnats

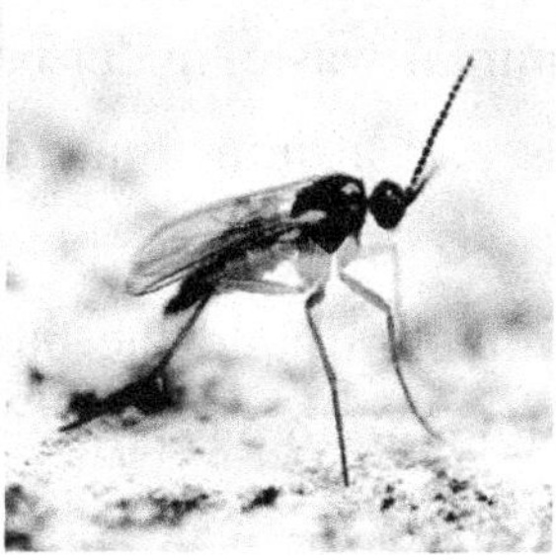

Order / Family: Diptera / Sciaridae and Mycetophilidae

Scientific Name: Various

Description:

Fungus Gnats are very small flies 1/32-to-7/16-inch long slender to robust, long-legged and mosquito-like. They are usually black, brown or yellowish. The dark winged Fungus Gnat (Sciaridae) has smoke-coloured wings, and the other Fungus Gnats (Mycetophiidae) have spots on their wings. The larvae are inconspicuous, although they have well developed heads, 11 to 12 body segments, and a ventral bottom lobe on their last abdominal segment.

Biology:

The dark winged Fungus Gnats moult four times as a larva. The eggs hatch in a few days and the larvae moult five times in 6 – 8 days. They pupate in the ground and adults emerge in three days.

Habitat and Economic Importance:

The larval food consists of moist decaying organic matter and fungi growing in the soil. These Gnats breed in various settings, such as rotting wood, animal waste, under bark, in over-watered plants and their saucers and atrium. Flat roof – structures and accumulated droppings in bird cages are potential breeding sites. Accumulations of old mulch around the building provide ideal habitat. They are small enough to pass through typical structural screening; adults are active at dusk are usually found near breeding sites. During the day, they seek out dark moist places to rest. They are attracted to light and often are seen at windows.

Common Site of Occurrence:

Fungus gnats breed in rotting wood, animal waste, under bark, overwatered plants, in potted plant saucers, atriums, growing media, foliage, compost, and wet mulch piles. Flat roof structures and dropping in bird cages and bird droppings accumulation indoors are potential breeding sites.

Management:

As there are hundreds of small fly species, the fly must first be identified before any course of action can be taken. The key to eliminating a Gnat problem is the removal of breeding sites. During the inspection, one needs to identify any site or condition which supports fungal growth. If adults are found around windows and doors, look outside for potential breeding sites, and remember that adults may be attracted to exterior lights. Please do also check for indoor potted plants as well, which can be a source of the problem.

Once the breeding site is located, affected materials should be removed or dried out. Interior plants should be over-watered but allowed to droop before watering. Insect light traps can be used to significantly reduce the indoor populations of these gnats. If fungal growth is found in any of the surfaces of the structure, suitable measures to be undertaken for preventing the moisture by leaking tap or water source, removal of fungal growth on the structure by using a suitable disinfectant chemical having fungicidal property. After the breeding site is eliminated, a spray application of residual insecticide should be used to kill the remaining adults.

34.

Furniture Carpet Beetle

Order/ Family: Coleoptera / Dermestidae
Scientific Name: Anthrenus flavipes LeConte

Description:

Adult Furniture Carpet Beetles are 1/16- to 1/8 inch long and in shape, covered with spots of yellow, white, and black scales. The underside of the body is white. The larvae are about 3/16 inch long, stubby-looking, and clothed with bands of stiff, erect brown hair. The body is wide in the middle but tapers at both ends. Three thick tufts of hair arise from either side of the body near the rear end.

Biology:

Females lay about 30-100 eggs in areas where the larvae are able to feed. The eggs hatch in 9- 19 days, and the larvae develop from 70 - 94 days before pupating in their last larval skin. The adults emerge from the pupal skin and may live for another 60 days. Development from egg to adult is 90 days - 2 years.

Habitat and Economic Importance:

This insect gets the name from its habit of feeding on the hair padding, feathers, or woollen upholstery of furniture and also feeds on wool, hair, fur, feathers, bristles, horn, tortoise shell, silk, insects, dried carcasses, dried cheese, dried blood, glue, and book bindings. It damages linen, cotton, jute, softwood, and leather when these items are soiled. Outdoors, the adults feed on plant pollen.

Common Site of Occurrence:

Carpet beetles are often found around the edges of rugs and carpets. Carpet beetles can also thrive on lint, hair, and debris accumulating under baseboards and inside floor vents and ducts.

Management:

Correct identification of the pest is the initial step in Carpet Beetle control, followed by a thorough investigation to identify and eliminate the sources of infestation. All products containing natural materials should be examined, paying particular attention to stored clothing, carpets, and food products. Sanitation, including removal and/or treatment of infested material, is the first step in controlling the problem. Carpets, rugs, and clothing should be brushed and cleaned.

Application of a residual insecticide like pyrethroids to cracks and crevices and the immediate area around an infestation is recommended. In cases where the infestation is associated with a wall void or other inaccessible space, a WP formulation that provides better residual activity should be used. The exterior of the structure inspected for plants that are touching the structures should either be removed or treated with the appropriately labelled product. During warmer months, perimeter foundation treatment with wettable powder or microencapsulated formulation is recommended. Carpet Beetle infestations are persistent and often require several follow-up visits in order to resolve the problem.

35.

German Cockroach

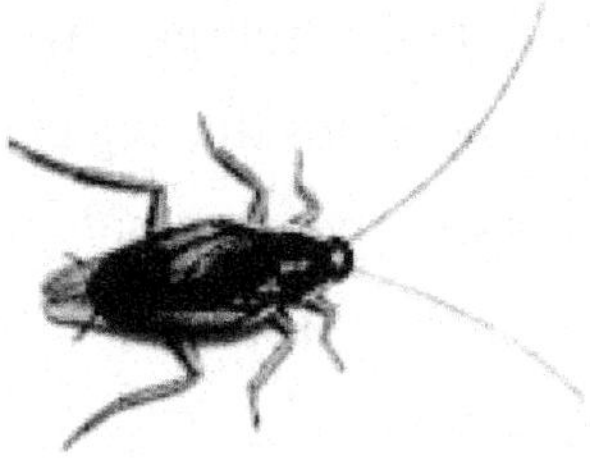

Order / family: blattodea/ blattidae

Scientific name: Blattella Germanica

Description:

German cockroaches are ½ to 5/8 inches long when mature, light brown to tan and have fully developed wings. The pronotum has two dark parallel bars on it. The adult males are somewhat narrower than the females when viewed from below. The nymphs, 1/8 inch long when they emerge from the capsule, are almost uniformly dark except for a light tan area on the back of the second and third segments. As they develop, the light tan area becomes larger until, as mature names, they have two parallel black bars separated by a light area. The purse-shaped egg capsule of the German cockroach is light brown in colour, ¼ to 3/8 inch long and typically has 15 to 20 eggs per site.

Biology:

During the lifetime, female German cockroach produces 4 to 8 eggs capsules, each of which contains 30 to 40 eggs. The female carries the egg capsule partially within her abdomen until just before the nymphs are ready to emerge. Approximately one to two days before hatching, she drops the egg capsule in a protected area. If the egg capsule is dropped prematurely, the developing roaches inside die of dehydration. Nymphs moult six to seven times before becoming adults. This requires about 103 days; Thus allowing 3 to 4 generations per year. Adults live 100 to 200 days. The established German cockroach population consist of approximately 75% nymph.

Habitat and Economic Importance:

German cockroaches are the most common household insect. These typically infects kitchens and bathrooms but will live anywhere inside heated structures in which there is food, water and harbourage. They are rarely found outdoors and then only during warm weather. German cockroaches gain entry into structures in grocery bags, cardboard boxes, drink cartons, equipment such as moved refrigerators, toasters, and microwaves, etc. Cockroaches feed on all types of human food, as well as pet food, toothpaste, soap, glue, etc. German cockroaches are active at night, leaving their harbourage to find food and water. They remain hidden in the dark, secluded harborage areas, for example, under cupboards, behind cabinets, in wall voids and around motor housings in appliances where they spend 75% of their time. At most, only one third of the population forages at night. Observation of foraging cockroaches during the day is a good indication that there is a tremendous population. Cockroaches congregate in harbourage sites; But as the population increases, overcrowding forces some of them to relocate.

Common Site of Occurrence:

German cockroaches are the most common species of cockroaches found in the households and commercial establishments in India. They have the ability to live anywhere where they get food water and harbourage but are most commonly found in the Kitchens or the locations where the food and food materials are stored. Major source of entry of German Cockroaches into the buildings / establishments / household is through grocery bags, cardboard boxes, electrical equipment's like refrigerators, microwave ovens if infested or any other kind of materials movement. You can find them in household kitchen, refrigerators, storage cupboards, Restaurant kitchens, Hotel Kitchens, Bathrooms, Hospitals storage cupboards, canteens, offices storage areas, educational institution storage areas, retails racks, storages etc locations.

Management:

Because German cockroaches are typically brought into structures, potentially infested products should be closely inspected. Cockroach Monitoring Traps are available to help pinpoint sources of cockroach infestation and to monitor areas about which occupants have complained about infestations but cannot visually be detected. Sticky traps are not intended for control but rather to guide and evaluate control efforts as part of the inspection process. Visual inspections can be conducted using a flashlight and flushing insecticides like pyrethroids or propoxur can be used as sprays to flush cockroaches from their harbourages.

An effective cockroach management program depends on good sanitation to eliminate the food, water, and harbourage they need for survival. Clean-up to reduce cockroaches in the home and office environments must focus mainly on the residue in and around coffee machines, microwave ovens, stores, refrigerators, trash cans, furniture, and areas where exposed food is stored. In addition, it is critical to reduce clutter as cockroaches like to hide in stock boxes, cartoons and stored paper and cardboard materials, especially in dark, damp locations near food.

Vacuum cleaning can be used to physically remove exposed cockroaches. Infested equipment may have to be disassembled and may have to be treated for complete population removal before further use. Permanent reduction of cockroach population can be achieved by sealing gaps/crevices to eliminate harborage. The most important cracks to eliminate include those at which sinks and fixtures are mounted, around crown molding hand corner guards, and where shelves and cabinets meet walls and door frames.

Containerized paste, dry and Gel baits have become very popular in the industry and are very effective products for managing cockroaches. To maximize effectiveness, paste, gel and dry baits should be applied with a syringe like dispensing tool in many small tabs or sports that are close to harbourage sites. Large, plastic bait containers should be placed as close as possible to the dark concealed spots where cockroaches are living, preferably adjacent to edges and corners.

Large populations of cockroaches can be reduced or eliminated with the careful application of spray and dust in sensitive locations. However, these products cannot be applied inside electrical appliances or computers. Many insecticides breakdown rapidly within the moist, cool locations these roaches prefer, so an appropriate formulation for the environmental conditions must be selected. Exterior barrier treatments with microencapsulated or wettable powder formulations might be effective.

36.

Grain Mites

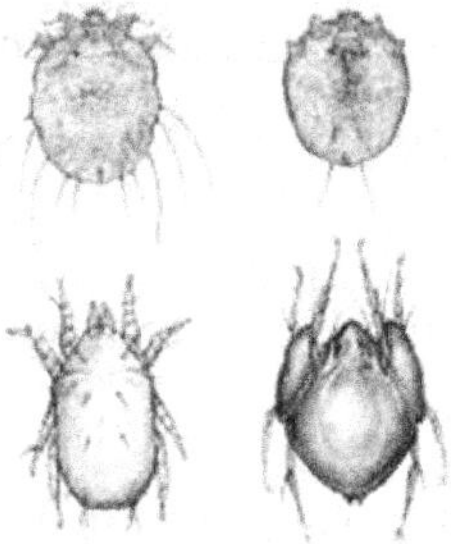

Phylum: Arthropoda

Class: Arachnida

Order: Acari

Scientific Name: Acarus siro (L.)

Description:

Grain mites are pests that can feed on a variety of processed or finely powdered grains, wheat germ, yeast, cheese, powdered milk, flour, or mold spores. Under hot, humid conditions, very large populations of these tiny creatures can develop rapidly. Grain mites are known to produce allergens on the food products on which they are feeding. They prefer to feed on dry, spicy products made up of black gram powder with spices included. They are a threat to the papad industry.

Habitat and Economic Importance :

Grain mites are not insects and are less than 1mm in size. They belong to Class Arachnida, characterized by the presence of 4 pairs of walking legs, unlike insects. The body segments are grouped into anterior prosoma, composed of 6 segments bearing 4 pairs of walking legs, and a hind or posterior region opisthosoma of 13 segments without appendages. These body regions are merged to give a single rounded appearance. They do not have antennae, but in some cases, the 1st pair of legs may be sensory in function. Compound eyes are absent, but simple eyes are in many pairs. Female lays up to 800 eggs on the surface of the material. The life cycle from egg to adult takes only about two weeks at normal room temperatures. Overcrowding in heavily infested products will force mites to move off in search of other food sources. This "spillover" can produce a fine dust-like layer of these mites on a shelf, product bag, or counter. This "mite dust" is often the first sign of trouble and usually means there are one or more heavily infested items not far away.

Common Site of Occurrence:

Grain mites are known to produce allergens on the food products on which they are feeding. They prefer to feed on dry spicy products made up of black gram powder with spices included. They are a threat to the papad industry. They can typically been seen in the warehouses, also in occasions of snacks or cereals infested in the household storage areas.

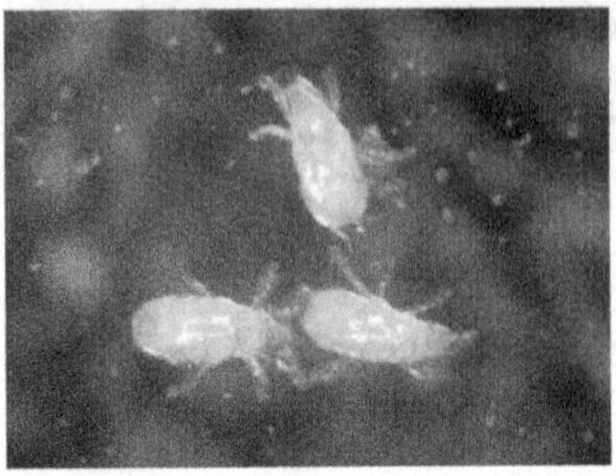

Management:

A thorough search must be made to locate infested packages so that they can be thrown away. All of the items on the shelf should be wiped off to remove and kill wandering mites before they enter another container. A rag with hot soapy water can be used to wipe up and clean out the mites on flat surfaces.

Cracks and crevices on counters or in shelving can be cleaned with the crevice tool of a vacuum cleaner to remove both mites and fine particles of processed foods that can keep the infestation under check. Immature stages of these mites can enter a resting stage that will allow them to survive for several weeks when conditions are not favourable. Thorough sanitation will help to remove as many of these as possible. To prevent re-infestation, pyrethroid liquid or dust formulations can be applied to cracks, crevices, surfaces and other potential harborage areas. Product applications are supplemental to the cleanup of infested material and should never be used as the sole means of control.

37.

Granary Weevil / Wheat Weevil

Order : Coleoptera

Family : Curculionidae

Scientific Name: Sitophilus granarius

Description:

Granary Weevil are similar to Maize Weevil and Rice Weevil. Adults are elongate, oblong, chestnut brown coloured, moderately shining beetles about 3 to 4 mm long, possessing a well-defined snout. Antennae geniculate with a large, compact club. The hind pair of membranous wings are modified or absent and hence cannot fly.

Biology:

The females may lay eggs between 30 and 250 in number; their eggs are deposited in a hole, excavated in the grain and plugged with a gelatinous secretion. Generally, there is a preference for the larger grains for oviposition. Eggs hatch in about 4 to 6 days; the small, white, legless larva remains in the grain, larval period about 25 to 30 days, single larva consumes about 28 mg of wheat during development and pupates inside the grain. Pupal period 10 to 16 days. When the development is complete, the adult chews its way to the grain surface life cycle 25 to 34days. Adults can live up to 8 months after emerging. One pair of weevils may produce up to 6,000 offspring per year.

Habitat and Economic Importance :

This is a primary pest of cereals, causing hollowing out of grains. As adults are devoid of membranous wings (hind pair), they are more confined to stored grain and do not attack crops in the field. Both adults and larvae are destructive and the adult life span is around eight months.

Common Site of Occurrence:

This insect is a primary pest in the temperate countries and cant withstand cold climates. They destroy and feed on Wheat and Cereals. These insects is evolved along with humans and can be a common stored pest insect in your household food storage areas as well if the infested grains or cereals are bought home. Found in warehouses and Cereal storage locations

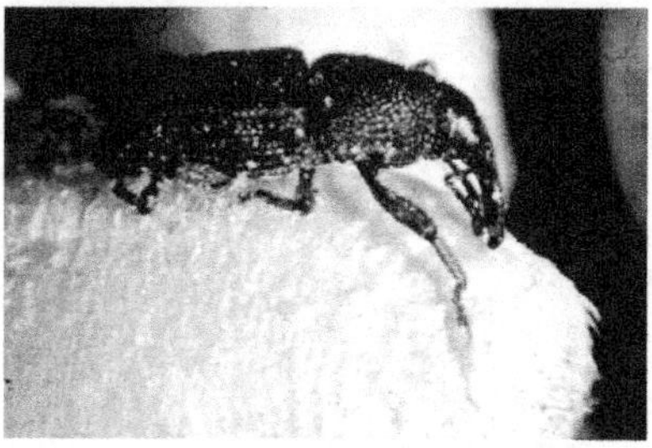

Management:

Sanitation and inspection are the keys to preventing the infestation. Grains should be stored in preferably metallic (cardboard, even fortified, is easily drilled through by the weevil) containers with tight lids in a refrigerator or a freezer and should be purchased in small quantities. If any suspicion has arisen, the grains should be checked for infestation of adult insects or holes in the grain kernels. Another method is to immerse them in water. If they float to the surface, it is a good indication of an infestation. Even if identified early, the disposal may be the only effective solution.

To provide control of exposed adults, residual insecticides can be applied to cabinets, closets and shelves where infested stored grains are found. All cracks and crevices are also to be treated along with any other potential harborage areas. Granary Weevil infestations in large storage facilities are successfully controlled with increased sanitation and bulk fumigation of the grain.

38.

Honey Bee

Order: Hymenoptera

Family: Apidae

Scientific Name: Apis Mellifera

Description:

Honey bees are one of the most beneficial insects in Agriculture. India has close to 800 bee species and "Over 95% of research on bees is on the Mellifera," as per Honey Bee Lab at the National Centre for Biological Sciences (NCBS), Bengaluru. Over a decade ago, Director Centre for Pollination Studies had assessed the economic value of pollination loss on six-seven crop productions — the annual loss could be around ₹5,400 crore.

In India, the different types of Bees include, Rock Bee, *Apis dorsata* (Apidae), the Indian Hive Bee, *Apis cerana indica* (Apidae), The Little Bee, *Apis florea* (Apidae), European or Italian Bee, *Apis mellifera* (Apidae), Dammer Bee or Stingless Bee, *Melipona irridipennis* (Meliporidae). Beekeeping in India has been mentioned in ancient Vedas and Buddhist scriptures. Rock paintings of Mesolithic era found in Madhya Pradesh depict honey collection activities. Scientific methods of beekeeping, however, started only in the late 19th century.

Biology:

The Queen Bee is the only fertile female and produces all the eggs for the colony and is capable of producing 1000 to 2000 eggs per day. Drones serve only to fertilize the Queens and they are driven off after they have served that purpose. Workers live 5 – 7 weeks during the summer. Honey Bees are social insects that live in the colony or high with as many as 20,000 to 80,000 workers.

Habitat and Economic Importance :

The workers collect nectar and pollen from plants, inadvertently pollinating flowers and allowing plants to produce fruit. They also produce honey and fashion honeycomb from the wax they secrete. The queen and all the bee larvae are fed and cared for by the young workers; older workers gather the pollen and nectar for the colony. Honey bees are not naturally aggressive; however, if the colony is threatened, they will sting. Honey bees swarm when the queen begins to fail, or the colony is too large swarms often are seen on a tree branch and when this occurs, the bees are not aggressive and the swarm lasts for 24 to 48 hours, moving to a sheltered environment such as a hollow tree beehive hollow wall or attic.

Common Site of Occurrence:

Honey Bee is an essential insect which is a need for the environmental ecosystem and its presence is a needful requirement and a major pollinator of the plants to product fruits. Honey bee are typically seen in an undisturbed or not much of human activity areas. As the colony size can be large to huge they prefer tree trunks, hollow tree, hollow wall, attics of high rise buildings.

Management:

Honey Bee colony is established within the wall of a structure: there are two control options, removing the living colony or killing them (*killing not recommended*). If the decision is to remove a living colony, it is best to work with an experienced beekeeper who can help safely remove the bees. Swarms are easily removed by an experienced beekeeper, removal of a colony within a structure may require several weeks but building occupants are rarely willing to wait that long because native honeybee colonies are rapidly declining.

39.

House Fly

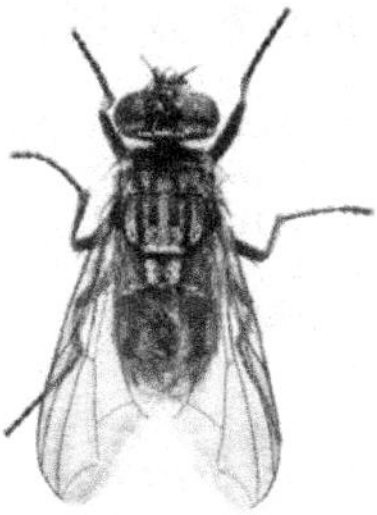

Order/Family: Diptera/Muscidae

Scientific name: Musca Domestica Linnaeus

Description:

House Flies are 1/8 – ¼ inch long. They are dull grey with four dark stripes on the back of the thorax. They have two wings; the fourth longitudinal wing vein has a sharp upward turn. The head is dominated by large red-brown compound eyes, surrounded by a light gold stripe. Short antennae emerge from between the eyes, and they have sponging mouth parts. Mature house fly larvae or maggots spindle shaped and creamy white. House Fly larvae are ¼ to 3/8 inch long .

Biology:

Female flies lay their eggs in clusters of 75 to 150 eggs in rotting, fermenting, organic matter including animal manure, accumulated grass clippings, garbage, spilled animal feeds and contaminated soil. A female may lay more than 500 eggs in a lifetime. The eggs hatch within a day and the young larvae burrow into breeding medium and complete development in three days to several weeks depending on temperature and quality of food materials. Larvae migrate to drier portions of the breeding medium to pupate for three days to four weeks before emerging as adults. Under optimum conditions, House Flies can complete their entire life cycles in less than seven days.

Habitat and Economic Importance:

The adult flies may migrate to un-infested areas up to 20 miles away, but most stay within one or two miles of the breeding site. Adult house flies have a general appetite, feeding on food ranging from the excrement of human food. They feed on liquids but can eat some solid foods by liquefying it with the help of digestive tract fluids. During the day, house flies rest less than five feet above the ground and at night, they rest above this height. House flies have been associated with many health-related diseases and thus are a significant health concern.

Common Site of Occurrence:

As the house flies have an appetite of feeding on range from human food to the excrement / garbage their presence is fairly wide spread. Their preference is seen to rest on the hanging objects. They are found in a restaurant, garbage heap, office canteens, retails, hospitals, hotels, manufacturing units, religious places, fairs, trade shops and any other places where there is a trace of food left uncleaned or spillage.

Management:

The initial inspection should focus on the identification of the fly (adult / larva) causing the problem and location of all resting and larval development sites. The adults often rest in breeding areas; it is helpful to inspect at night. Sanitation along with source reduction, is the most important step in housefly control because it eliminates larval breeding sites. When successful, it significantly reduces the need for pesticides applications. If trash cans are the problem, property owners must be instructed to empty and clean them at least weekly intervals.

Mechanical control measures include insect proof garbage containers, self–closing doors, screening, air curtains. Sticky traps and other devices are also available to reduce adult fly populations indoors and outdoors.

Any insecticide applications should be directed at adults because sanitation and removal are the best control measures for larval breeding sites. Products used for fly control include baits, aerosols and residual insecticides. Most residual applications are made to the daytime and evening resting sites. Ultra-low volume and aerosol applications should be made when the adults are most active and there is the least risk of drift, surface contamination, and human or animal exposure.

40.

House Mouse

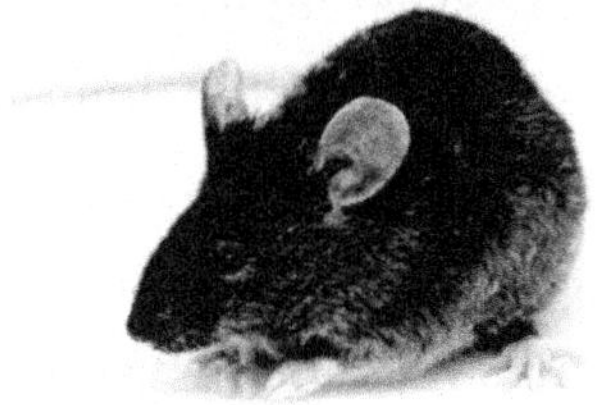

Order / Family: Rodentia/ Muridae

Scientific Name: Mus musculus Linnaeus

Description:

The House-Mouse is the most common and economically important mammal living in close association with humans. The house mouse is Gray in colour and it weighs ½ to 1 ounce. The body is 3 to 4 inches long and the tail is three to four inches long the muzzle is pointed, the ears of large, the ice and body are small. Typically, the house mouse is slightly smaller than deer mice. Adult droppings are one/8 to 1/4 inch long and rod shaped with pointed ends.

Biology:

The female reaches sexual maturity in 35 days and average 8 litters per year. Populations build up very rapidly. They typically produce the largest litters in the spring, depending on climatic conditions and begin to breed at five to six weeks of age. The lifespan is one year.

Habitat and Economic Importance:

House Mice are good climbers who jump 12 inches high and down from 8 feet. They squeeze through holes and gaps wider than a quarter inch. They are very social in their behaviour, very inquisitive about things in their environment, and readily exclude anything new. They prefer to nest in dark, secluded areas where there is little chance of disturbance, where nesting materials, such as paper, cardboard, attic insulation, cotton, etc., are readily available. They nibble on food, preferring items such as seeds and cereals. They feed at dusk and just before dawn.

There can be major health risks associated with these creatures.

Common Site of Occurrence:

Though the name is house mouse but this species is not restricted only to households. They prefer to nest in dark and very secluded place where there is little or less chance of disturbance. They choose areas like storage of papers, cardboards, cotton, storage in attics etc places in household and commercial establishments like offices, educational institutions, hotels, retails etc. places

Management:

Droppings and rub marks are signs of their presence but in general, is less noticeable than those produced by rats. Entry holes should be sealed with quarter inch hardware cloth, sheet metal, or metal wool. Pet foods and other food products should not be stored in accessible areas, such as garages and water resources.

Within structures, house mice can be controlled effectively by using baited and unpainted snap traps and glue boards. Traps can be baited with nesting materials, such as cotton, string, dental floss, etc. or fruit, vegetables and seeds. Their inquisitive nature makes them easy to trap; Thus, periodically moving the traps increase success.

Several anticoagulant rodenticides are available as pellets, packets and blocks. When water sources are limited, liquid anticoagulant baits are very effective but must be handled in the same way as dry baits. The attractiveness of liquid baits can be increased by formulating the rodenticide with low concentration fruit juice. Tracking powders are also available, which contain either an acute single dose toxicant or anticoagulant active ingredients.

41.

House Sparrow

Order / family: Passeriformes / passeridae

Scientific Name: Passer domesticus.

Description:

House Sparrows are approximately 6 inches long. Their colour varies with sex, males having a black patch under the beak and the cheeks, top of the head is grey white. In the winter, the black area is hidden by the crave feather tips. Females and young sparrows are dull brown with dirty white breast and brow.

Biology:

The eggs range in colour from white to light green, do light blue and have numerous dark spots. They average five to six eggs per clutch and require 10 to 14 days to incubate. Two to three broods per year or up to 21 young per year are produced. After hatching, the almost featherless young birds are totally dependent on the parent birds for food and warmth.

Habitat and Economic Importance:

Sparrows prefer to nest in protected man-made or natural areas. Building ledges, openings in structures, gutters, signs, light fixtures, birdhouses, bridges, common electrical power lines and transformers are used to construct nests. Sparrows also displace other birds (e.g., wrens, robins, and purple martins) from their nest. Both sexes construct rather large and flimsy nests from straw, cross, feathers, strips of paper, spring and other debris. HouseSparrows frequently nest in kitchen, bathroom, oven, and dryer vents.

Sparrows also enter food plants, warehouses, department stores and malls, where they often contaminate food products or other merchandise. Their droppings can contain a variety of disease-causing bacteria, fungi, nematodes, etc. Numerous blood-feeding parasitic mites associated with sparrows also bite humans.

Common Site of Occurrence:

These little birds though are nesting outside buildings generally but prefer to nest in protected areas or natural areas where there is good protection. They choose places like building ledgers, openings in structures, gutters, light fixtures, bird houses, bridges, electrical power lines, transformers. Sparrows also destroy nests of other birds like wrens, purple martins, robins etc to make it as their own nest for brooding. They frequently visit inside of the structures and favourable conditions they can establish the nest inside the building structures as well in kitchens, dry vents, roof gaps etc locations. Very commonly seen in Large commercial spaces like warehouses, Malls, Retails, Transportation Terminals, educational institutions etc locations.

Management:

Both non-chemical and chemical means of eliminating bird problems exist. Sanitation, which removes the food, water, and protection the birds seek, is an important first step. Exclusion involves structural modifications which roosting and nesting. Options include plastic and metal spikes, plastic netting, pin and monofilament line slash wire, and the use of repellent gels and pastes. Sound devices might be effective but are considered unsuitable for residential environments. Trapping often is low, labour intensive, and effective only if the birds are destroyed instead of released. Shooting is the final alternative for obvious reasons has limited application and regulatory approval implications.

42.

Khapra Beetle

Order: Coleoptera

Family: Dermestidae

Scientific Name: Trogoderma granarium

Description:

The Khapra Beetle, Trogoderma granarium , is one of the most feared stored-product pests. In fact, it has been described as one of the 100 worst invasive species. Established infestations are difficult to control because of the beetle's ability to live without food for long periods of time and to survive on foods of low moisture content. These beetles tend to crawl into tiny cracks and crevices and remain there for long periods, making them relatively tolerant to many surface insecticides and fumigants. Therefore, it is important to prevent the Khapra Beetle's introduction into un-infested areas. Products infested: Cereals, groundnuts, groundnut cake, pulses and spices. Primary pests, a very serious pest of wheat.

Khapra larvae, in particular, the diapausing ones, are one of the most tolerant among stored products insects to insecticides, including fumigants.

Biology:

Small oval beetles, 1.5 to 3 mm in length, densely covered with scales, pale reddish-brown in colour. Elytra cover the abdomen and are without striae, have red-brown markings: Head deflexed and retractile. A median ocellus is present between the compound eyes. Antennae with the distinct club. The females are about twice the size of the males. Larva is Straw-coloured with numerous reddish-brown coloured tufts of setae, arranged segmentally. Antennae is 3-segmented and the pupa are adecticous exarate.

Habitat and Economic Importance:

The larvae do not usually feed on the materials of animal origin but prefer cereals and oilseeds, most dreaded among the stored products insects. Originated in India, the adults are short lived (9 to 12 days); they do not feed and do not fly. About 60 to 70 eggs per female are laid on the food material. The larvae with a profuse number of setae are easily transported to different places. The larvae have the habit of crowding over into crevices of buildings, bins and other storage structures. Under unfavourable conditions, many mature larvae enter a resting state or diapause. The life cycle takes 25 to 28 days under optimum conditions.

Common Site of Occurrence:

Though virtually cosmopolitan this species is favoured by hot dry climates. Primary pests, a very serious pest of wheat. Largely found in the storage areas like Warehouses, Wheat processing areas, carried infested grains can be source for spreading this to retails and households as well.

Management:

The obvious signs of a Khapra Beetle infestation are the larvae and cast skins. However, the larvae look very similar to those of other relatively unimportant Trogoderma species, as well as some Carpet Beetles. Larvae and adults are best identified by microscopic examination. Detection methods include examination of cracks and crevices and inspecting behind the panelling on walls and under timbers, tanks, shelves, etc. Larvae are most likely to be seen just before dusk since they tend to be more active at that time. Some fumigants give control at high dosages, even though this Beetle is more resistant to fumigants than most stored product pests. High concentrations of fumigant must be maintained over the fumigation period to allow penetration into all cracks and crevices. In an eradication program, both fumigants and surface sprays are used in combination with preventive measures, e.g., good sanitation practices and exclusion.

43.

Lady Bug

Order: Coleoptera
Family: Coccinellidae
Scientific Name: Various

Description:

Ladybugs also called as Ladybird Beetles, are a very beneficial group. They are natural enemies of many insects, especially aphids and other sap feeders. A single Lady Beetle may eat as many as 5,000 aphids in its lifetime. Adults have characteristic convex, hemispherical to oval shaped bodies that can be yellow, pink, orange, red, or black and usually are marked with distinct spots. This is a type of warning colouration to discourage other animals that may try to eat them. The adults have 3 tarsal segments, the part of the leg farthest from the body, a characteristic feature that distinguishes them from beetles of similar size and shape, which have 4 tarsal segments.

Biology:

Most coccinellids have round to elliptical, dome-shaped bodies with six short legs. Depending on the species, they can have spots, stripes, or no markings at all. Seven-spotted coccinellids are red or orange with three spots on each side and one in the middle; they have a black head with white patches on each side. Eggs are orange and are laid on end in single or multiple groups. Once hatched, the larvae eat about 350 to 400 aphids in the 2 weeks it takes them to become fully grown. As they grow, the larvae shed their skin several times. Once they are a certain size, the larvae stop eating and attach themselves to a plant leaf or stem. They then enter their next developmental stage, called the pupa. Ladybugs can survive for up to nine months by living off their stored reserves.

Habitat and Economic Importance:

These insects are beneficial because the larvae and adults both eat a variety of outdoor ornamental pests such as aphids, mealybugs and other soft bodied insects and their eggs. In the early winter, they gain shelter under rocks, bark, leaves, and timbers to hibernate over the winter period. In this period, they prefer the house or the occupied structures. Adults are attracted to light and once inside the structure, they are seen in light fixtures or window sills.

Common Site of Occurrence:

These insects are beneficial insects in the social environment and their presence in the nature is good as the adults feed on the variety of outdoor ornamental pests. They prefer to nest in areas such as beneath rocks, barks, leaves and landscape timbers. They are predominantly outdoor pests of Lawns and green areas only and not found inside the structures unless brought in by the materials of the lawns.

Management:

Prevention is one of the best control methods. Use a fast-acting synthetic pyrethroid to help prevent bugs and similar pests from entering homes. For best results, the exterior walls have to be sprayed from the ground up, and once Ladybugs have become established in the home, the easiest way to remove them is with a vacuum cleaner.

It is also recommended to spray around attic vents, windows, doors, under-fascia lips, siding and any other possible points of entry. Sealing out the cracks and around utility service openings, fixing broken window screens and door jams, plugging cracks in the foundation or roof, as well as similar exclusion-type activities, will help prevent these beetles.

Periodic inspection of exterior plants for aphids, scale and other soft bodied insects will help to anticipate Lady Bug infestations. Preferable, these plants should be treated with systemic insecticide or a topical formulation to control the prey of Ladybug and minimize their interest in the structure.

44.

Lesser Grain Borer

Order: Coleoptera

Family: Bostrychidae

Scientific Name: Rhyzopertha dominica

Description:

The adult body is cylindrical, about 2 to 2.3 mm long, dark reddish-brown in colour, head deflexed and more or less concealed from above. Wings with well-defined rows of punctures, prothorax and elytra covered with strongly curved sitae. Antennae inserted laterally, 10 segmented with a large, loose, 3-segmented club.

Habitat and Economic Importance :

Lesser Grain Borers mainly attack wheat, corn, rice and millet.

Both the larvae and adults are primary pests. They bore irregularly shaped holes into a whole, undamaged kernels and the larvae, immature stages, may develop inside the grain. Larval and adult feeding in and on grain kernels may leave only dust and thin brown shells.

It is a primary pest of cereals, including paddy and amongst the most voracious and destructive pests. It reaches its greatest potential as a pest in the warm, drier areas when grain stocks are left undisturbed for long periods. It has also been recorded from wood, bamboo, cork, dried roots and books.

Biology:

Most of the newly-hatched larvae chew into kernels and complete their entire development there. Larva Scarabaeiform, 0.3 mm long when hatched and measures 2.8 mm when matured, with legs having single claw, the first instar larval, last abdominal segment bears a median posteriorly curved dorsal spine.

Adults have a long lifespan. 250 to 400 eggs per female, laid on the surface or among the cereal grain over a period of 3 to 6 weeks, either singly or in batches. Newly laid eggs are white and glistening, then turn rosy and opaque, hatch in 6 to 14 days. Larvae penetrate the grain after their 2nd ecdysis and many larvae feed on the powdery food material. Several beetles may be found on one grain and the larvae pupate inside the grain. The adults have the habit of frass making before laying eggs. This frass consists of food material chewed off but not eaten. Development is possible in grain with 8% moisture content. The life cycle takes about 25 days at 34oC R. dominica has developed very high levels of resistance (x100 and above) to phosphine in many regions.

Common Site of Occurrence:

These are Major pest of cereals and infest products like cereals, cereal flour, split pulses and even paddy. It is most voracious destructive pest in cereals. This reaches to high population when the grains are kept undisturbed for long periods. Major pest of Storage areas like warehouses, mills and other storage facility. Can also be a household problem if the stored grains infested and kept undisturbed for long periods.

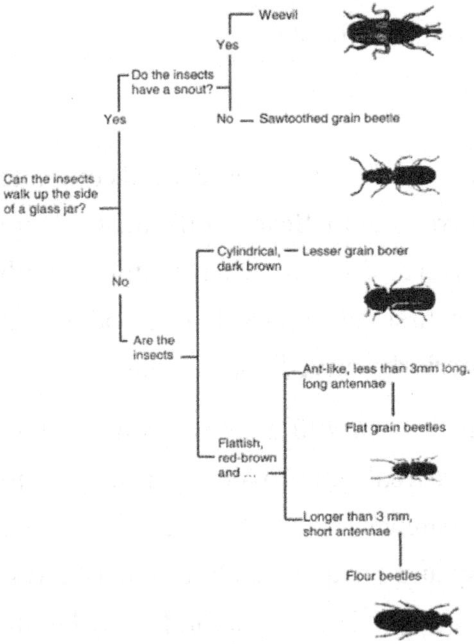

Management:

Prevention is always the most economical and efficient method of controlling these pests. Once they are distributed within the grain mass, fumigation is the only method of relieving the problem. Removing spilt grain from around storage facilities can reduce Lesser Grain Borer populations. Damaged grain is distinctive (as detailed above). Infested grain may give off a sweet smell. Sieving a sample of grain is the simplest detection method, but it will only detect adults and larvae that are outside the grain. Where the infrastructure exists, freezing for several days and heating for 24 hours have proved to be effective control methods for stored product pests. The insecticides Chlorpyrifos-methyl and pirimiphos-methyl, while effective against most stored grain insect pests, are not very effective against the Lesser Grain Borer. The fumigant phosphine can be used in sealed storage facilities.

What is the difference between a Beetle &Weevil?

Beetle	Weevil
Mouth parts typically chewing type.	Mouth parts chewing type but modified into snout like (nose like proboscis like) structure.
Both pairs of wings present, fore (Forward) wing elytrate, which at rest in straight suture when at rest.	Hind wing absent, fore wing elytrate, immovably united
Antennae capitate (club, knob like) / serrate (angled on one side) /Tameilate	Antennae clavate (clublike).
Tarsi (final segment in the leg of insect) 3 to 5 segmented	Tarsi (final segment, furthest from the body) 4 segmented.
Larvae oligopod (Thoracic legs are well developed)	Larvae apodous (without appendages for locomotion)

45.

Little Black Ant

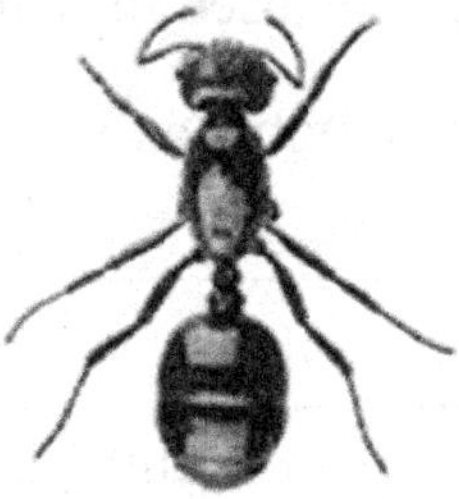

Order: Hymnootera

Family: Formicidae

Scientific name: Monomorium minimum

Description:

The thorax of black antlacks spines and the petiole has two nodes. The abdomen is tipped by a small stinger. Each antenna ends in a three segmented club. Little blank ant workers are one by 1/16thinch long. These same characteristics also describe Pharaoh ants as well; the difference, however is the characteristic which distinguishes these two species is the jet black colour of the little black ant.

Biology:

Colonies have several queens and grow rapidly in a fairly short period of time. The winged reproductive typically swarm in late spring or early winter. Although small, little black ants will fight larger ant species over food resources, especially when they're protecting aphids, a favourite food that satisfies their sweet tooth.

Habitat and Economic Importance:

Little black ants are found nesting beneath stones in lawns and in areas that lack vegetation. Nests are easily located as they form small craters of fine soil at their entrances. These ants also nest within rotting wood and behind the woodwork or masonry of structures. Indoors, they can be found under the edge of carpeting and within old termite galleries and wall voids. Little black ants prefer to feed on a variety of food, including honeydew produced by Aphids, meats and greases. They feed on other insects indoors; they feed on greases and sweet foods.

Common Site of Occurrence:

These insects are found nesting outdoors like rotting woods, underneath stones, in lawns, behind woodwork and masonry structures. These are found commonly in areas in outdoors where there is lack of vegetation. Indoors they are found where there is availability of greases and sweet foods and prefer under carpeting, old termite galleries, wall voids, door frames gaps etc places.

Management:

All cracks and crevices within exterior walls which provide access to voids and interior areas should be sealed. All debris from the exterior of the structure and other items on the ground where these ants can nest should be removed. Firewood should be stored above the ground. Little black ants are best controlled by locating their nest and treating them with a residual insecticide.

Exterior Nests should be drenched with the liquid formulation. Nest within wall voids is more easily treated by spray application or a dust formulation. Since these ants typically nest outdoors, forage indoors and live in small colonies, it is rarely necessary to apply barriers treatment to the foundations. Baits are very effective in controlling colonies and should be placed where Ant trails have been established and in locations where the ants have been sighted. Baits used indoors should be placed such that they are inaccessible to children and pets. Sweet baits are the most effective.

46.

Millipede

Class: Diplopda

Family: Various

Scientific Name: Various

Description:

In India, this insect comprises over 270 species or subspecies in at least 90 genera, 25 families and 11 orders. Anoplodesmus sp., Orthomorpha coarctata and Xenobolus carnifex are all commonly seen and are capable of composting waste. From an evolutionary point, millipedes were among the first to inhabit land during the Ordovician period (approximately 450 million years ago). Mainly herbivores and mostly detritus feeders (organisms that feed on dead plant or animal matter), it is believed that Millipedes may have played a major role in soil formation and nutrient cycling during the Paleozoic era. Although the word millipede is derived from Latin words mille-thousand and ped-foot, no single species discovered so far have that many legs. On average, legs present may be from 11 to 375 pairs.

Biology:

In size, millipedes range from a body length of 1.4 mm to 30 cm long. When hatching from eggs, they do not have many segments and legs. They show what is known as an anamorphic development; that is, new segments and legs are added after every moult. Over time they undergo 7 to 10 moults, during which the number of legs and segments increase. Many species reach sexual maturity within two years, but some require four or five years to complete development. The adult can live several years, breakdown of decaying organic matter during feeding, coupled with fecal matter providing a surface for microbial action, makes the millipede a valuable composting agent.

Habitat and Economic Importance:

Millipedes are attracted to dark, cool, moist environments that are rich in organic matter, such as compost piles, heavily mulched shrub or flower beds, rotting logs, or the soil under logs and stones. They usually go unnoticed because they live in these relatively hidden habitats. Millipedes are scavengers, feeding primarily on decomposing vegetation but occasionally damage soft-stemmed plants in gardens. Major nuisance problems usually occur when the conditions become too hot and dry and the millipedes move in search of moisture. Large numbers of them congregate in wet areas such as around in-ground swimming pools where they may crawl (and fall) into the water. Millipedes may also migrate in the early winter, presumably in search of hibernating sites. All activity result in millipedes invading crawl spaces, basements and other areas of buildings. Common points of entry include door thresholds, expansion joints, and through the voids of concrete block walls.

Common Site of Occurrence:

These are the pests commonly found outdoors in places such as decaying organic matter, trash, rotting wood, leaf litter, unclean drains etc places. However they enter into the structure if their natural habitat is challenged and forced to move where they find indoors or during the mating season in search of mate and if the weather favors with high moisture content in indoors. These are found in basements, porch areas, lawn areas, unattended greenery areas, bathrooms, drains, and moisture containing walls, ducts, wood areas inside the structures.

Management:

Recommended to seal cracks and openings in the outside foundation wall and around the sills of doors and basement windows. Expansion joints and gaps should also be sealed along the bottom of basement walls on the interior to reduce the entry of pests and moisture from outdoors. The most effective, long-term measure for reducing the entry of millipedes (and many other pests) is to reduce excess moisture and hiding places, especially near the foundation.

Once indoors, millipedes end up in kitchens, living rooms, and soon die from a lack of moisture. Applications of insecticide outdoors may help to reduce inward invasion of these and other pests. Treated along entryways, around crawl space doors, foundation vents and utility openings, and up underneath siding. Insecticides applied along the interior foundation walls of damp crawl spaces and unfinished basements may help temporarily.

47.

Norway Rat

Order/ Family: Rodentia / Muridae
Scientific Name:Rattus norvegicus (Berkenhout)

Description:

The Norway rat is the largest of the commensal rodents. The head and body are seven to 10 inches long and the tail is an additional 6 to 8 inches. It has a stocky body and weighs 7 to 18 ounces. The fur is coarse, shaggy, and brown with some black hairs. The muscle is blunt, eyes and ears are small, and the tail, which is bicoloured, is shorter than the head and body combined. Norway rat droppings are up to 3/4 inch long with blunt ends.

Biology:

Adults are sexually mature in two to five months. Females produce three to six litters per year, each averaging 7 to 8 young. Adults live from 6 to 12 months. They have poor sight but keen senses of smell, taste, hearing, and touch.

Habitat and Economic Importance:

Rats are nocturnal. They are shy about new objects and very cautious when things change in an environment and along with their established runs. Norway rats prefer to nest in Burrows in the soil, for example, under sidewalks and concrete pads, riverbanks, railroad track beds, next to buildings, in low ground cover, etc. The burrows typically have one main entry hole and at least one escape hole. The rats easily enter buildings through half an inch and larger gaps. In buildings, they prefer to ncst in the lower levels of the building, for example, crawl space, basement, loading dock and sewers. They prefer foods such as meat, fish, and cereals and require a separate non-food water source. Their foraging ranges 100 to 150 feet from their nest. Rats are associated with various diseases and occasionally bite. The plague is one and leptospirosis is vectored by rats and, thus, is a great disease of concern. This disease is acquired by eating food and drinking water which are contaminated with infected rat urine. Rat also caused significant structural damage and product destruction.

Common Site of Occurrence:

These are pests which are commensal in nature and can have their harbourage in any location of availability of food and shelter. They prefer to nest in burrows in the soil near the side walks, concrete pads, banks of water bodies like stream, river etc., rail track beds, next to building soil areas, mid / mezzanine structures with soil. They normal have their nest within 100 to 150 feet of available food source. These are common pests in Storage facility, commercial establishments, offices, restaurants, hotels, lodgings, households, public amenities, educational institutions, residential units etc locations.

Management:

The keys to a successful program of rodent control are identification, sanitation, elimination of harbourage, and rodent proofing. The inspection should identify signs of infestation, e.g., nibling marks, droppings, tracks, rub marks, damaged goods, etc. Sanitation consists of removing food, water and materials which provide harbourage. Storage should be at least 12 inches off the floor and 18 inches away from the wall.

Rats can be trapped using glue boards and snap traps placed along with walls and near vertical runs where the rats travel. Traps can be baited with fish, meat and cereal, or they can be left un baited. Because rats are scary of new things in the environment, it may be helpful to leave the traps on set for a while.

Exclusion is a critical aspect of rodent control. Since rats can squeeze through a half inch gap, anything larger must be sealed. Since they can chew holes, ceiling smaller holes should be considered. Sheet metal, cement, quarter inch hardware cloth, etc., all the materials of choice which can be used for service. Door sweeps should be installed around gaps on doors, windows, and other openings.

Several anticoagulant rodenticides are available as pirates, packets, and blocks. When baiting indoors, these products should be placed in tamper-resistant bait stations and in areas that are then accessible to children and pets. Outdoors, palpates can be placed dipped in the burrow using a long-handled spoon and then the burrow should be closed. If packets and blocks are used in Burrows, there is a risk that the rat will push the product outside the burrow. When water sources are limited, liquid anticoagulant baits are very effective but must be handled in the same way as dry baits. Tracking powders are also available, which contain either an acute single dose toxicant or anticoagulant active ingredients and typically are used to dust Burrows and runs. Using gas in outdoor Burrows is an option but is risky particularly if the borrow extends under or into a structure.

48.

Odorous House Ant

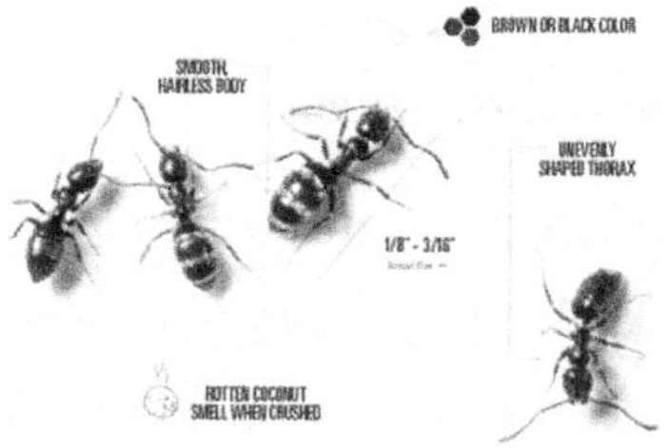

Order: Hymnootera

Family: Formicidae

Scientific name: Tapinoma sessile

Description:

Odorous House Ant workers are 1/16thto 1/18thinch long and colouration is brown to black. The thorax has no spines and the petiole has one segment which, when viewed from above, its hidden by the rest of the abdomen. There is a slit at the tip of the abdomen instead of a circular band of hairs. The best identifying characteristic is the rancid butter smell produced when the ants are crushed. Hence, their name Odorous House Ant.

Biology:

These ants swarm to mate from early May through mid-July; they also made within the nest forming new colonies by budding off from the original colony. Approximately 10,000 workers and several Queens in the colony & each lay one egg per day. Developmental time from egg to adult is 30 - 90 days. During winter, it may take six to seven months to develop. There are several generations per year.

Habitat and Economic Importance:

Odorous House Ants often nest outdoors beneath stones and logs and within the nest of larger ants. They also nest indoors in voids of the floor, around heat sources such as hot water pipes, heaters, heating equipment, cracks and crevices, kitchen sinks and cabinets. They travel in trails and prefer sweets, although they eat almost any household food. They usually invade structures during rainy periods after honeydew on plants has washed off.

Common Site of Occurrence:

These ants are commonly found in the outdoors and indoors as well. They travel in trails and prefer sweets but can eat almost any household food. They usually invade the structure during the rainy season as the source of food will be washed off outside. In outdoors they prefer nesting beneath stones and wood logs, indoors they prefer wall voids, floor voids, around heat sources such as hot water pipes, stove, heaters, crevices around sink and cabinets. They prefer to travel in trails only.

Management:

All cracks and crevices within the exterior wall that provides access to interior areas should be sealed. All debris from the exterior of the structure and other items on the ground where these ants' nest is to be cleared. Although odorous house ants prefer sweets, they are not easily controlled with sweetened baits. Baits should be placed where Ant trails have been established or have been sighted. Bait placement should be such that they are inaccessible to children and pets.

Sweet baits are effective in summer and in winter, protein baits are more effective. The most effective control is accomplished by following ant trails locating the nest and treating the nest with the residual insecticide. The Exterior nest should be drenched with a liquid formulation. Nest within wall voids is to be treated with a spray application or a dust formulation. When the nest cannot be located, a barrier spray treatment using a non-repellant product is the most effective treatment. Microencapsulated or wettable powder repellent formulation can be used, which not only repel and kill foraging workers they also can trap them inside

49.

Oriental Cockroach

Order/family: blattodea/ blattidae

Scientific name: Blatta orientialis Linnaeus

Description:

The male Oriental cockroach is one-inch long and female is 1-1/4 inches long when mature, and red-brown to black instruct. In males, the wings cover 75% of the abdomen; In females, they are reduced to small wing pads. The other nymphs are light brown but become increasingly darker with each moult. The purse shaped egg capsule is dark red-brown but becomes black with age, 3/8 inch long.

Biology:

The female Oriental cockroach produces an average of eight egg capsules, each of which contains 16 eggs. The capsules are dropped off, using secretions from mouth, glued to the protected locations such as cracks and crevices near food sources. Their nymphs moult about 10 times before becoming adults. Depending on the temperature, this requires 206 - 800 days. Adult females live from 34 - 181 days and males from 112 - 160 days.

Habitat and Economic Importance:

Oriental Cockroaches are not common pests in most homes. They usually are found in sewers and commercial facilities, like prisons, restaurants, hospitals, offices and apartment buildings. They often are found in bathtubs and sinks because they lack the small pods on their tarsi, commonly found on other cockroaches, which allow them to crawl up smooth surfaces. Outdoors, even in cold weather, they are found in groundcovers, stones, leaf litter and other debris. Oriental cockroaches prefer to feed on starchy foods but eat various other items, including decaying organic matter. Oriental Cockroaches produce a very characteristic pungent "cockroach" odour.

Common Site of Occurrence:

These insects are not the common household pest and are rarely found in most homes. They are seen abundant in drainages, sewer lines and commercial facilities associated with large drain lines and kitchen drains. They are also found in areas of damp, secluded places such as basements, water meter boxes, sewer holes, etc damp associated places. They are good climbers on smooth surfaces and are seen near bathtubs and sinks. Outdoors they are found underneath stones, leaf litter and other damp debris. They are found in hotels, kitchens, municipal drains, Residential societies drains, hospitals, grocery stores, offices and commercial kitchens.

Management:

Cockroaches often are brought into the facilities via equipment and storage boxes. Thus, potentially infested products that are bought into structures should be closely inspected. Cockroach monitoring traps are available to help pinpoint sources of the cockroach infestation. Good flushing agent treatments like propoxur can be used for spray treatment to flush cockroaches from their harbourages.

An effective cockroach management program depends on good sanitation in order to eliminate the food, water and harbourage they need for survival. It is critical to reduce clutter as large cockroaches like to hide in stacked boxes, cartoons, rolled carpet and stored paper and cardboard materials, especially in the dark, damp locations. Vacuum cleaning can be used to physically remove exposed cockroaches & their egg case.

The most important cracks at which sinks and fixtures are mounted to the wall and/or floor, around all types of plumbing, baseboard moulding, vessels and cabinet meet walls, door frames, and cracks on or near food preparation surfaces. Cockroach access routes should be sealed with waterproof filler or paste. Basement floor drains should be protected with screens or basket inserts which must be cleaned regularly. To maximize effectiveness, paste, gel and dry baits should be applied with a syringe. Large, plastic bait containers should be placed as close as possible to the dark concealed spots where cockroaches are living, preferably adjacent to edges and corners. A large population of cockroaches can be reduced or eliminated with careful application of sprays and dusts, followed by application of paste, gel and dry baits. Many insecticides breakdown rapidly within the moist, hot locations these cockroaches prefer, so an appropriate formulation for the environmental conditions must be selected.

50.
Pharaoh Ant

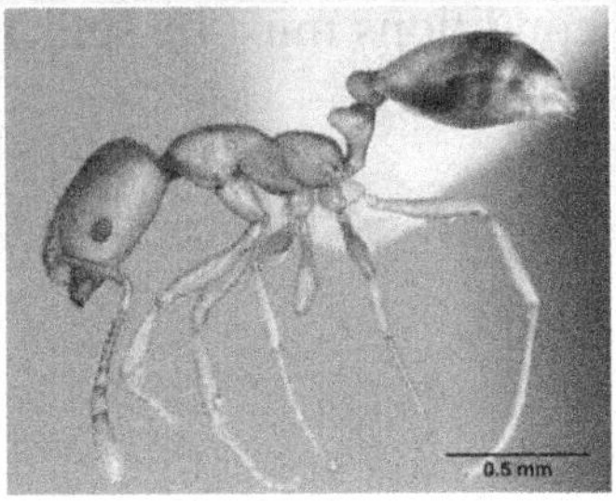

Order: Hymnoptera
Family: Formicidae

Scientfic Name: Monomorium pharaonis

Description:

Pharaoh ants get their name from the mistaken belief that they were one of the plagues of Egypt during the time of Pharaohs. Pharaoh ants are very small, with workers about 1/16th inch long, colouration ranges from yellow to light brown. The Thorax region lacks spines and the petiole has two nodes. These ants can be distinguished from other ants by three segmented club and the end of their antenna

Biology:

There may be hundreds of thousands of ants within a colony, there are usually many queens in a colony and they co-exist amicably. The queens can be replaced so colonies will survive indefinitely. New colonies are typically formed by budding and may be encouraged by disturbing nests. A female produces 350 to 400 eggs in her lifetime. Indoors these ants develop all around the year. Workers live approximately 9 to 10 weeks and queen for four to 12 months.

Habitat and Economic Importance:

Pharaoh Ants are widely distributed and nest outdoors and are at times a crop pest. They are a major problem in homes and institutions such as hospitals, hotels, apartment complexes. They nest within warm hard to reach locations like subfloor areas, wall & electric sockets, attics, cracks and crevices and behind baseboard and furniture. Pharaoh ants eat dead and live insects but seem to prefer moths. They also feed on sugar syrup, fruit juices, jellies and cakes. These ants are an especially important species of pests within hospitals where they have been found infesting the dressings on the patient's wound, feeding on secretions from newborn infants in IV tubes etc.

Common Site of Occurrence:

Pharaoh ants are major pests in indoors as well as outdoors. They nest within the warm and hard to reach places such as walls, sub floor areas, wall sockets, attics, cracks and crevices, behind baseboards and furniture's. Very commonly found insects in Hospitals feeding on dressing of wounds, secretions of IV tubes etc. They are also pest of house holds, manufacturing units, apartment complex, hotels etc commercial establishments.

Management:

Pharaoh Ants nest in a wide variety of locations within structures, making it difficult to control. They do not necessarily follow specific trails of food sources and are difficult to trace to their nest. However, it is useful to look near sources of water and food. A successful Pharaoh Ant control program involves intensive baiting using products that contain IGR and baits should be placed in areas where ants are expected to be active. The more placements, the better the result. They also should be placed behind outlets and switch plates in areas where wires pass through walls and within cracks and crevices.

The most effective control measures involve a thorough insecticide treatment so all insects constituting the infestation will be exposed to the toxicant within a short span of time. This is best achieved by baiting, but conventional insecticide treatments may be necessary to affect a rapid control of ants in some situations or establish perimeter treatments to protect areas from ant immigration. However, conventional treatments alone are of limited value for the control of Pharaoh's ant nests.

The elimination of Pharaoh Ants from a structure might require much more time than a month or so. Residual application of insecticides should not be used in the control of Pharaoh Ants. These application effects stop less than 5% of the workers in a colony but stress the colony and cause the colony to split, leading to colonization of new areas. Spraying gives a little reduction in the existing population.

51. Pigeon

Order/ family: Columbiformes/ Columbidae

Scientific Name: Columba livia domistica

Description:

Pigeon is approximately 13 inches long, with the colou varying from white to black. However, the Rock Dove is characterized by the dark grey head with an iridescent sheen, light grey back, and rings with two dark bands. It has a stocky body with short legs and a neck and a small head. During the flight, the tip of the tail is usually square and black in colour.

Biology:

The white eggs are laid 1 - 2 at a time and require approximately

18 days to incubate. Two to five broods are produced per year or up to 10 young per year. The eggs are incubated by both parents. After hatching, the almost featherless young birds are totally dependent on the parent birds for warmth and food. For the first five days, the newly hatched birds fed a Milky substance, "pigeon milk," which is produced in the crop of the parent birds. During the next five days, more water and grain is incorporated into the "milk", and finally, they are fed only grain and water. The young pigeons leave the nest approximately one month after they hatch. Pigeons live from 4 to 12 years.

Habitat and Economic Importance:

Pigeons have become the most serious bird pest associated with buildings. They nest in a variety of protected locations, such as in the underside of bridges, building ledges, open buildings, roofs, air conditioners, etc. The loosely constructed nests typically consist of sticks, stems, leaves and other debris. Nests that are reused often become solid with the accumulation of toppings and debris.

In rural settings, pigeons typically feed on seeds, grain and fruit. They find areas with spilled silage, such as grain elevators, railroad yards, and mills, and all of which are very attractive feeding sites. In urban areas, pigeons feed on handouts, garbage, vegetable matter and insects.

In contrast to many other bird species, pigeons prefer flat and smooth services, such as rooftops, for resting and feeding.

Pigeons gather in flocks that use the same resting and feeding areas. Feeding usually occurs no more than a few miles from the resting site. Their very acidic droppings can cause significant damage to the equipment, painted building surface is, marble, limestone, etc. Dropping also contaminates unprocessed grain and processed food and can contain a variety of disease-causing bacteria, fungi, nematodes, etc. Pigeons serve as reservoirs for several viral encephalitic diseases. Many of the parasitic mites associated with the pigeons also bite humans.

Common Site of Occurrence:

Pigeons are the most serious bird problems in the buildings, particularly in the household high rise structures. They rest in variety of structures such as building ledges, roofs, air condition ducts, signages, underside of bridges, etc man made locations structures. They roost and nest where there is easily and plenty of food available. They nest during all seasons when condition permits.

Management:

Both non chemical and chemical means of eliminating word problems exist. Sanitation, i.e., removing the food, water, and protection the birds seek, is an important first step. Exclusion involves structural modifications which resting and nesting. Options include plastic and metal spikes, plastic netting, pin and monofilament line/ wire, and the use of repellent gels and pastes. Sound devices might be effective but are considered unsuitable for residential environments. Trapping often is slow, labour intensive, and effective only if the birds are destroyed instead of released. Shooting or final alternative for obvious reasons has limited application. There are bird repellent gels/pastes available which can be applied to the Bat roosting areas, but these will have a limitation on the period of effectiveness not being too long and also repeated application would be required of the same.

52.

Psocids / Booklice

Order/ family: Psocoptera / Various
Scientific name: Various

Description:

Psocids, commonly called Book Lice or Paper Lice, are not lice at all. In fact, they look more like termite workers but are much smaller. They are 1/32 to ¼ inch long, pale, and the wings, if present, are usually held in a root like position over the body when resting. The antennae are long and thread like.

Biology:

Psocids undergo gradual metamorphosis, i.e., the immature forms resemble the adult forms. The developmental time of egg to adult ranges from 24 - 110 days, though this varies greatly with species and environmental conditions. Low relative humidity hinders development or causes death due to desiccation.

Habitat and Economic Importance:

Indoors, Psocids seek out damp, warm, undisturbed areas, such as inside books on shelves, mouldy foodstuffs in pantry, and stacks of stored paper goods, they are found in old window and door casings where moisture accumulates and mould grows. In new structures, they are attracted to the moisture in fresh plaster. In warehouses, the most common complaints about the pest are associated with wooden pellets and palletized materials. Outdoors, they seek out protected areas, e.g., under tree bark or leaves, mulch, bird and rodent nests, etc. High moisture is essential to their survival. Book lice feed on various substances, including moulds, fungi, starches and dead insects. Certain species fed on the starchy flue and paste that binds books and wallpaper to surfaces.

Common Site of Occurrence:

Psocids are attracted towards moisture and mould growth areas. They are found where there is high moisture content. In indoors they are found in damp, warm undisturbed areas such as book shelves, mouldy foodstuffs in pantry stacks or storage, Paper goods or storage having moisture. In outdoor areas they seek protected areas like under tree barks or leaves, bird and rodent nests. In warehouses it is commonly seen associated with wooden pellets and materials on the pellets.

Management:

Inspection for Book Lice is difficult because their presence often is unnoticed; they are difficult to detect without the aid of a hand lens. The most successful control measure for Book lice is to reduce the humidity in infested areas, which dries out infested goods and eliminates mould, mildew and other food sources. Ventilation increases moisture evaporation, thus, eliminating the humid conditions that support microscopic mould and fungi growth on packaging materials, grains, and other products. ULV applications kill exposed Book lice. Cracks and crevice treatment is effective in areas such as book shelves or storage racks where Book lice activity is observed. Products which contain silica aerogel and other types of desiccants are effective. In extreme cases, particularly in sensitive accounts, such as food and drug manufacturing facilities, fumigation of the site may be necessary.

53.

Rice Moth / Indian Meal Moth

Order: Lepidoptera

Family: Pyralidae

Scientific Name: Plodia interpuctella

Description:

Rice moths are flying insects that can be spotted flying around in a zig-zag fashion in homes. They have a wing expanse of about 3/4". When they rest, they keep their wings on their bodies rather than spreading them out like a butterfly. Fore wing uniformly coloured pale, grayish brown without spots but with the veins slightly darkened. The legs are long and slender. Rice Moth are one of the most common stored products pests found in homes, food processing plants, grain storage and processing facilities. Products infested: Rice, sorghum, millet, wheat, cocoa, oil seeds (peanuts), copra, milled products and some spices. They can damage rice flour mills, causing aggregation of the infested material with their silken webbing.

Biology:

Females lay 150 to 200 eggs at night on larval food over 1-to-18-day period. The matured larva is about 15 mm long and 3 mm wide. The cuticle of this species is exceptionally white compared to other species associated with stored products. Setae without pigmented spots on the cuticle, the larval head is pale brown in colour. Pupa typically about 7.5 to 9 mm long, with short median, parallel, elevated, longitudinal lines. The average life cycle requires 4 to 5 weeks under optimum conditions, usually 4 to 6 generations in a year, depending on the food supply and temperature. Mature larvae move out of the food material to the cracks and crevices to pupate. The adults are more active at dusk and are attracted to light. The adults are short lived about 1 to 2 weeks and do not feed. The damage caused by these insects feeding on food is compounded by the presence of the repulsive webbed mat.

Common Site of Occurrence:

This insect are seen majorly destroying cereals, ground nuts and dry fruits. It avoids direct light hence they seek shelter in dark or lesser light areas in the storage / warehouse. It can be a common stored pest insect in your household food storage areas as well if the infested grains or cereals are bought home. Found in warehouses and Cereal storage locations

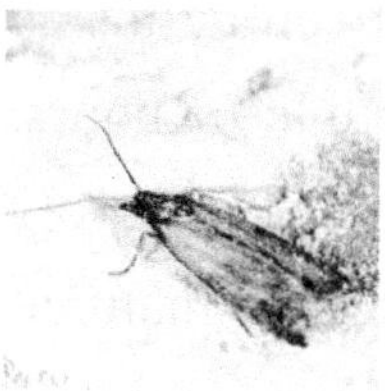  

Management:

These insects find themselves around dry foods and packet foods. They enter homes from open windows or crack on the walls. The first sign of rice moths flying inside homes is a clear sign of rice moth infestation. There is a good chance that they might have already laid their eggs on food. The second sign is the most obvious one, clusters of silk webbings on food, especially inside the rice and flour containers. The larvae of rice moths create these clusters. Inside these clusters, there are larvae poop and the dead skin that they shed. The third sign of an infestation is molding on your dry food. The larvae of rice moths make these molds while they are eating rice and other dried foods.

First, it is advised to check food storage, food containers, and food in each of the repositories. If the food storage containers are thin, then rice moths can quickly drill through them to lay their eggs inside. If they are loosely packed, then they can also sneak in. So, a thorough check must be done inside every food storage jar, including Tupperware, for the signs of rice moth infestation. Clean kitchen cabinets and pantry storage thoroughly, and it is better if a vacuum cleaner is used. Any holes or cracks in the kitchen and across the home is an entry point for the Adult Rice Moth. Treat the infested food to get rid of the larvae and the dirt. If it is rice, then it can be washed with warm water. If it is flour, cereals, dried fruits, pasta, and pet foods, it can be inside the refrigerator or spread out in the sun for a couple of hours. Both heat and cold kill the larvae of rice moths. Pyrethroids spray can be applied in all cracks and crevices surfaces for the matured larvae and pupae destroying. For commercial storage of grains or for large warehouse storage, an infestation can be controlled by a good sanitation process and fumigation is the best method to kill any infestations.

54.

Rice Weevil & Maize Weevil

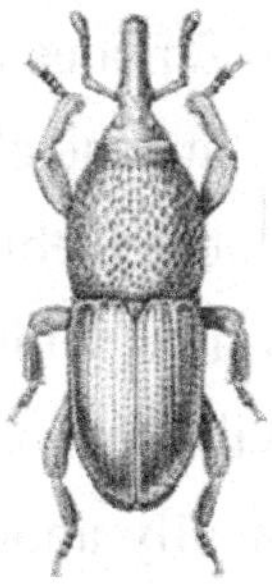

Order: Coleoptera

Family: Curculionidae

Scientific Name: 1. Sitophilus oryzae = Rice Weevil

2. Sitophilus zeamais = Maize Weevil

Description:

The Maize Weevil is slightly larger than the Rice Weevil and has more distinct coloured spots on the forewings. It is a stronger flier than the Rice Weevil. The habits and life cycle are similar to the Rice Weevil. Rice and maize weevils are usually found in grain storage facilities or processing plants, infesting wheat, oats, barley, rice, and corn. Although not often found in the home, they are sometimes found infesting beans, birdseed, sunflower seeds, dried corn, and to a lesser degree, macaroni and spaghetti. These weevils do not bite, nor do they damage wood or furniture.

Biology:

Reddish-brown to dark, blackish, tan coloured weevils, 2.1 to 2.8 mm long. They are distinguished from all other beetles with a well-defined snout (rostrum, beak, proboscis). The rostrum more slender and thinner in females and short, stouter in males. Antennae elbowed is geniculate and clubbed. Thorax, longer than wide constricted near apex, sides feebly curved, divergent towards the base. Wings deeply striated with four yellowish or reddish spots. S. oryzae (Rice Weevil) are smaller than S. zeamais (Maize Weevil). Mature larva, typically apodous, 2.5 to 2.75 mm long, pearly white, fleshy and very thick bodied, dorsally semicircular and ventrally flat. Pupa Exarate, having no mandibles, uniformly white, when first formed and gradually darkens. Head with 2-prominent spines towards vertex, a group of two small spines and two spiracles on each side above the eyes.

Habitat and Economic Importance:

Both the species have similar life cycles. Rarely more than one larva develops in a single grain because of cannibalism. The developed adults remain inside the grain for several days before emerging. Although there are no clear preferences for particular grains, the strain and varietal differences of grains affect the reproduction of Sitophilus species. S. oryzae is more common in small cereal grains and has a higher tolerance than S.zeamais. The adults generally avoid strong light and become very active when disturbed. Both the species can fly thus are capable of infesting the grain in the field itself before harvest. The flight activity is more pronounced in S.zeamais.

Under optimum conditions, 100-150 eggs are laid by the female over a period of many weeks. A peak in the egg laying activity occurs about three weeks after emergence from the pupa. Each egg is laid in a minute hole chewed in the grain by the female and is sealed inside the hole by a secretion. Eggs are shining white, opaque, ovoid top ear shaped, bottom broadly rounded, 0.68 to 0.80 mm long and 0.33 mm wide. The legless larva remains inside the grain, feeds and eventually pupates there. Single larva of S. oryzae consumes about 10 mg of grain and about 0.4 mg of wheat is consumed daily by an adult. When the development is completed, the adult bites its way out of the grain, leaving behind an emergence hole. Both adults and larvae are destructive. The adults are long living up to one year and lay eggs throughout, although more than 50% of eggs are laid in the first 4 to 5 weeks.

Common Site of Occurrence:

Primary pest of cereals which causes hollowing of grains. These are more prevalent in humid warm regions, maize grown areas in the field, local storage areas in field, warehouses and central storage locations of cereals.

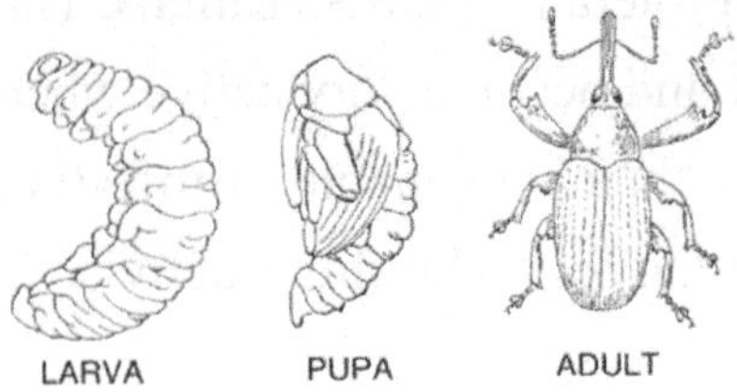

Management:

Rice Weevil adult presence indicates the presence of infested whole grains somewhere on the premises. Infested materials should be checked and discarded or subjected to extreme heat or cold treatment. Closets, shelves, cabinets and pantries should be thoroughly cleaned, after which a residual product can be applied to the surfaces as well as into cracks, crevices and other potential harborage areas. Large storage facilities that are infested with Rice Weevils need to increase sanitation and eliminate the infestation with proper fumigation methods.

55.

Roof Rat

Order/ family: Rodentia / Muridae

Scientific Name: Rattus rattus Linnaeus

Description:

The Roof Rat is a commensal rodent. The head and body are 6 to 8 inches long and the tail is an additional 7 to 10 inches. It has a slight body that weighs 5 to 9 ounces. The muscle is pointed, eyes and user large, and scaly tail, which is uniformly dark, is longer than the head and the body combined. Roof Rat droppings are up to half inch long and spindle shaped with pointed ends.

Biology:

Adults are sexually mature in two-five months. Females produce four to six litters per year, each averaging 6 to 8 young, the adults live from 9 to 12 months. They have poor sight but keen senses of smell, taste, hearing and touch.

Habitat and Economic Importance:

Rats are nocturnal. They are shy about new objects and very cautious when things change in an environment and along with their established runs. The burrows typically have one main entry hole and at least one escape hole. The rats easily enter buildings through half inch and larger gaps. In buildings, they prefer to nest in the lower levels of the building, for example, crawl space, basement, loading dock and sewers. They prefer foods such as meat, fish, and cereals and require a separate non-food water source. Their foraging ranges 100 to 150 feet from their nest. Rats are associated with various diseases and occasionally bite. Leptospirosis is vectored by rats and, thus, is a great disease of concern. This disease is acquired by eating food and drinking water which are contaminated with infected rat urine. Right also caused significant structural damage and product destruction.

Common Site of Occurrence:

These are nocturnal, live and make their shelters / harbourage areas in the above ground places. In outdoors they prefer to nest in trees and vegetation. Inside the structures they prefer high raise areas like fall ceiling, roof thatch, upper storage areas etc places. Their nest will be 100 to 150 feet from the area of activity. They are most common pests in the Office buildings, Hospitals, Educational institutions, Manufacturing plants or industries, Hotels and Lodgings of all categories, Restaurants, etc commercial establishments

Management:

A successful program of rodent control is with identification, sanitation, elimination of harbourage, and rodent proofing. The inspection should identify signs of infestation, e.g., nibbling marks, droppings, tracks, rub marks, damaged goods, etc. Sanitation consists of removing food, water and materials which provide harbourage.

Rats can be trapped using glue boards and snap traps placed along with walls and near vertical runs where the rats travel. Traps can be baited with fish, meat and cereal, or they can be left un-baited. Exclusion is a critical aspect of rodent control. Since rats can squeeze through a half inch gap, anything larger must be sealed. Since they can chew holes, ceiling smaller holes should be considered. Sheet metal, cement, quarter inch hardware cloth, expandable forms comma etc., all the materials used of choice which can be used for service. Door sweeps should be installed around gaps on doors, windows, and other openings.

Several anticoagulant rodenticides are available as pellets, packets and blocks. When water sources are limited, liquid anticoagulant baits are very effective but must be handled in the same way as dry baits. The attractiveness of liquid baits can be increased by formulating the rodenticide with low concentration fruit juice. Tracking powders are also available, which contain either an acute single dose toxicant or anticoagulant active ingredients.

56.

Saw-toothed Grain Beetle & Merchant Grain Beetle

Order: Coleoptera

Family: silvanidae

Scientific Name:

1. Oryzaephilus surinamensis= Saw-toothed Grain Beetle

2. Oryzaephilus mercator = Merchant Grain Beetle

Description:

Adult beetles are depressed, elongate and narrow, 2.5 to 3.5 mm long. Body dark, chestnut brown coloured, clothed with lighter pubescence. Prothorax - with six large teeth on each side and three prominent ridges on the dorsal surface, antennae moniliform with a compact club. Elytra completely cover the abdomen, each with 4 ridges or striae. The two species may be separated on the length of the temple, the region between the compound eyes and the posterior margin of the head. The posterior femora of the males are armed with a tooth, absent in females. ***O. surinamensis*** feed on cereals and milled cereals, dry coconut, oilseeds, oilseed cake etc. ***O. Mercator*** feed mainly oil seeds and derivatives, infests rice, rice bran, spices and dried fruits. Both the species are secondary pests of whole grains causing only superficial damage and the primary pests of milled products, copra and oil seed cake.

Biology:

O. *mercator* can fly and is less frequent. About 370 eggs per female are laid loosely in the food material over a period of 10 weeks, at the rate of 6 to 8 eggs per day. The eggs hatch in 3 to 5 days into slender, pale cream-coloured larvae, with two slightly darker bands on thoracic and abdominal segments. The larvae are active and move about freely, feeding upon the germ portion of the cereals. They construct a silken cocoon in which they pupate. Both adults and larvae are destructive. The adult life span is one to three years and the life cycle takes 27 to 35 days under optimum conditions.

Common Site of Occurrence:

These are insect pests of milled cereals, dry coconut, oilseed, oilseed cake etc oil seed derivatives. These are more prevalent in humid warm regions, local storage areas in field, warehouses and central storage locations of cereals.

Management:

In homes, infestations can be avoided by storing dried food products in sealed containers. To control already present infestations, the infested material needs to be identified and disposed of or frozen- as all life stages of the beetle can be killed by being frozen for six days. In food processing operations and warehouses, other means of control may be necessary and fumigation is a commonly used and effective control method in large-scale grain storage operations. Pesticide application may be needed for storage over six months. Fumigation is commonly used to control stored product pests in food and grain. The surfaces cracks and crevices and other potential harborage areas for these pests can be treated with residual insecticide sprays. Application of these insecticides will also help to control adults that crawl into storage areas serving as potential re-infestation threats. Sanitation is very important in all grain storage facilities to keep potential breeding sites to a minimum.

57. Scorpions

Order: scorpiones
Family: various
Scientific Name: various

Description:

Scorpions are one of the most dangerous invertebrate animals found in the world. The Red Indian Scorpion is a big threat to humans and responsible for many cases of killing a human using its venomous sting. Deathstalker species of Scorpion is the most dangerous species of scorpion in the world, found in desert habitats of the Thar Desert in India. Indian Red Scorpion from Buthidae family is amongst the world's most lethal and dangerous scorpion species. This species occurs in most of India and causes most of the human fatalities in India. Other species found in India are Indian Black Scorpion (Heterometrus Bengalensis), Giant Forest Scorpion (Heterometrus Swammerdam), Fat tail Scorpion (Androctonus), Euscorpiops Asthenurus.

Scorpion instances of poisoning with high morbidity and mortality is usually due to either excessive autonomic activity and cardiovascular toxic effects or neuromuscular toxic effects. Antivenin is the specific treatment for scorpion envenomation combined with supportive measures, including vasodilators drug in patients with cardiovascular toxic effects and benzodiazepines when there is neuromuscular involvement.

Biology:

They have eight legs and are easily recognized by a pair of grasping pincers and a narrow, segmented tail, often carried in a characteristic forward curve over the back and always ending with a stinger. Male and Female Scorpion grasp each other's pincers and move around in a "dance" where the male tries to maneuver the female onto his deposited sperm packet. Most scorpion species are nocturnal, finding shelter during the day in burrows, cracks in rocks and tree bark. Many species dig a shelter underneath stones a few centimeters long.

All known species give live birth and the female cares for the young as their exoskeletons harden, transporting them on her back. Nymphs undergo an average of 6 moults over a period of several months to 4 years before reaching maturity. Adults live from 1 year to 6 years and can survive up to six months without feeding.

Common Site of Occurrence:

These are found mostly in dry and deserted areas and are active at night. They hide beneath stones, tree barks, under rock and wood piles and in Masonry cracks in outdoors. They only enter the inside of a building or a structure for the sake of water and shelter and are seen in places such as bathrooms, crawlspaces, attics, dry store, foundations and in some cases inside the shoes left on the floor outdoors.

Management:

Scorpions that cause a nuisance outdoors can be controlled by removal of harborage areas, including trash pile stones, boards, firewood landscape timbers and debris. Entry into structures can be prevented by sealing and caulking gaps around siding windows, doors pipes and wires. Microencapsulated and wettable powder products which eliminate insects and other arthropods eaten by scorpions are the most effective formations that can be used in the outdoor habitats preferred by the scorpions. These products should be applied in a 3-to-10-foot band around the perimeter of the structure into harborage sites and around other potential entry points.

Indoor Scorpions can be controlled by sealing their harborage areas wherever possible and vacuuming to remove exposed Scorpions can be introduced in two wall whites' voids between floors epics and other potential hiding places.

58.

Silverfish

Order/family: Thysanura / lepismatidae

Scientific Name: Lepisma saccharina (Linnaeus)

Description:

Silverfish are primitive, wingless insects that become half-inch long when fully grown. They are covered with silvery scales that are flat and somewhat carrot-shaped. Three long, slender antennae-like appendages project from the end of the optimum, giving them the name bristle tails.

Biology:

The female lays 1-3 eggs per day in crevices or under objects. The female moults after laying a batch egg and sheds her skin as many as 50 times after becoming an adult. The eggs hatch in about 43 days at 72- 90-degree Fahrenheit. Under ideal conditions, they moult every two to three weeks becoming adults in three to four months. However, under poor conditions, this might require two to three years. These insects have a long life-span.

Habitat and Economic Importance:

Silverfish are tropical insects that easily adapt to the structural environment. They live in warm locations; they hide during the day and rest in tight cracks and crevices. They roam great distances looking for food, but once a food source is located, they remain closed until the supply is exhausted. They can be found throughout the structure, from the basement to individual floors to attics to shingles on the roof. They readily feed on books, cloth, and sometimes dried meat or dead insects; they seem to especially be fond of books, paper, glue and pastes.

Common Site of Occurrence:

These can be found in the indoors wherever there is warm and moist locations or conditions. They hide and breed in cracks and crevices. They are seen in all locations from ceiling to basement depending on the food source available. In outdoors beneath roof shingles, in mulch and cracks and crevices of the structure. They have an affinity to books, paper and book glue so it can be noticed commonly in the book storage and shelfing locations also.

Management:

Reducing moisture, lowering the temperature, removing infected items can eliminate local infestations. They can survive for weeks without food and water. Silverfish are easily controlled with careful and thorough applications of residual insecticide sprays and dusts. Care should be taken to treat wall voids, cracks and crevices, and other suspected harbourage areas thoroughly. The same careful application of residual insecticides used for good cockroach control is needed to effectively control silverfish. Outdoors, microencapsulated or wettable powder formulations should be used.

59.

Smoky Brown Cockroach

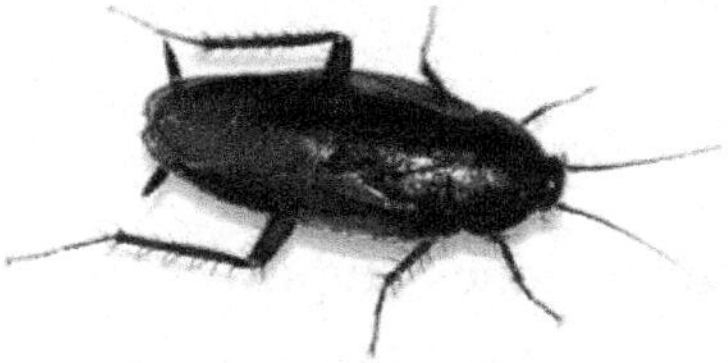

Scientific name: blattodea/ blattidae
Scientific Name: Periplaneta fuliginosa

Description:

Smokey brown cockroaches are 1 to 1¼ inches long when mature, uniformly dark brown to mahogany, and characterized by fully developed wings that completely cover their abdomens. The pronotum is black and the nymphs are red-brown. The younger ones have white markings on their backs and the first four or five segments of their antennae are white at the tip. The purse shaped egg capsule is dark brown to black; 3/8 inch long and typically has 10 - 14 eggs per side.

Biology:

The female smoky Brown Cockroach produces approximately 10 egg capsules, each of which contains about 20 eggs. Using secretions from our mouth, the female firmly attaches the egg cases to some surface or object covers them with surrounding materials. Nymphs emerge in 50 days and moult 10 to 12 times. Depending on the temperature, development from egg to adult requires 160 to 716 days. Adult females live an average of 218 days and males about 215 days.

Habitat and Economic Importance:

Smokey Brown Cockroaches typically are outdoor pests. Outdoors they are often found in wood piles, flower planters, palm trees, water oaks and vacant buildings. Indoors they seek warm, humid areas without air circulation, such as garages, attics and crawlspaces. Smokey brown cockroaches have also been found established in green houses. These cockroaches are strong Fliers and are attracted to light.

Common Site of Occurrence:

These are typical outdoor pests found near the vegetations like wood piles, palm trees, and also a pest found in the vacant unoccupied buildings. Indoors they seek warm and humid conditions without much of air circulation like attics, garages, crawlspaces etc. places. So we can see these species in large commercial establishments majorly & in residential societies.

Management:

Cockroaches often are brought into the facilities via equipment and storage boxes. Thus, potentially infested products that are bought into structures should be closely inspected. Cockroach Monitoring Traps are available to help pinpoint sources of cockroach infestation and to monitor areas that have complained, but infestations cannot be visually detected. Visual inspections can be conducted using a flashlight. Good Flushing agent treatment like propoxur can be used for spray treatment to flush cockroaches from their harbourages.

An effective cockroach management program depends on good sanitation in order to eliminate the food, water and harbourage they need for survival. It is critical to reduce clutter as large cockroaches like to hide in stacked boxes, cartoons, rolled carpeting and stored paper and cardboard materials, especially in the dark, damp locations. Vacuum cleaning can be used to physically remove exposed cockroaches & their egg case. cockroaches are particularly sensitive to drying, so it is important to reduce moisture by repairing leaks, improving drainage, and installing screened vents in order to increase air flow. Permanent reduction of cockroach populations can be achieved by waterproofing filter to eliminate harbourage and prevent entry into structures.

The most important cracks are those at which sinks and fixtures are mounted to the wall and/or floor, around all types of plumbing, baseboard moulding and cabinet meets walls and door frames, and cracks on or near food preparation surfaces. Cockroach access routes from wall voids into occupied spaces and around plumbing and electrical fixtures should be sealed with caulk or grout. Basement floor drains should be protected with screens or basket inserts which must be cleaned regularly. Containerized, paste, dry and Gel baits have become very popular in the industry and are very effective products for eliminating cockroaches. To maximize effectiveness, paste, gel and dry baits should be applied with a dispensing tool like syringes in many small tabs or sports that are close to harbourage sites. Large, plastic bait containers should be placed as close as possible to the dark concealed spots where cockroaches are living, preferably adjacent to edges and corners.

A large population of cockroaches can be reduced or eliminated with careful application of sprays and dusts, followed by application of paste, gel and dry baits. Many insecticides break down rapidly within moist, hot locations. These cockroaches prefer, so an appropriate formulation for the environmental conditions must be selected.

60.

Social Wasp

Order: Hymenoptera

Family: Vespidae

Scientific Name: Vespa mandarinia (Asian Giant Hornet)

Description:

A Wasp is any insect of the order Hymenoptera which is neither a bee nor an ant. Many wasps, can sting their insect prey. The most commonly known wasps, such as hornets or Asian giant hornet. With their powerful stings and conspicuous warning colouration, often in black and yellow, social wasps are more frequent ones. All species of social wasps construct their nests using some form of plant fiber, mostly wood pulp as the primary material, though this can be supplemented with mud, plant secretions, and secretions from the wasps themselves; multiple fibrous brood cells are constructed, arranged in a honeycombed pattern, and often surrounded by a larger protective envelope. Asian Giant Hornet has a body length of 45 mm to 75 mm long and a stinger of 6 mm, the hornet's head is a light shade of orange and its antennae are brown with a yellow-orange base.

Biology:

Social Wasps have large nests which contain three types of individuals or castes: Queens Workers and Males. The Males and Queens are produced in the colony in late summer, they mate and the fertilized queen hibernates in a protected site. In spring, the queen seeks an appropriate nesting site in which it builds a paper nest using chewed up wood fibers eggs are laid within the cells of the nest and the young larvae are fed bits of chewed meat or Insect parts by the queen and later by the workers.

Habitat and Economic Importance:

Social wasps are considered pests when they become excessively common or nest close to buildings. People are most often stung in late summer when wasp colonies stop breeding new workers; the existing workers search for sugary foods and are more likely to come into contact with humans.

Wasp nests made in or near houses, such as in roof spaces, can present a danger as the wasps may sting if people come close to them. Stings are usually painful rather than dangerous, but in rare cases, people may suffer life-threatening anaphylactic shock.

V. mandarinia is the only species of social wasp known to apply a scent to direct its colony to a food source. The hornet secretes the chemical from the sixth sternal gland; the ability to apply scents may have arisen because the Asian Giant Hornet relies heavily on Honey Bee colonies as its main food source.

Social Wasp used their nest for only one season. Unlike Bees, these aggressively defend their nests and can inflict multiple stings. They produce very large colonies, some containing as many as 30,000 individuals. These insects are considered to be beneficial because they feed their young a wide variety of insects. They become a nuisance, however, when they build nests in a structure, scavenge for food in recreational areas and in other places frequented by humans and seek hibernating sites with structures.

Common Site of Occurrence:

Normally Hornet species they prefer to build their nest in trees and on the sides of the buildings. They also build nest in wall voids. Yellow jacket species usually build their nest below ground areas protected locations like basements, stilts etc in the building. Paper wasps build their nests usually around the house in protected locations like attics, window silts corners, etc places. Normally found in more greenery around area in the commercial as well as residential locations.

Management:

Removing the nest will not resolve the problem because surviving Wasp will construct a new one. Pesticide formulations like synthetic pyrethroid are effective against social wasps. A light application is enough to eventually kill all the wasps as they pass through the opening.

The best strategy is to treat the nest at night when all the workers and queen are present. This tactic maximizes the effect of pesticide application by killing most of the Wasp. If treatment is made at night, shining a light directly on the nest should be avoided. Daytime treatments can also be effective. Once the nest is removed, the nest attachment area should be treated with residual synthetic pyrethroid.

61.

Spice Beetle; Drug Store Beetle

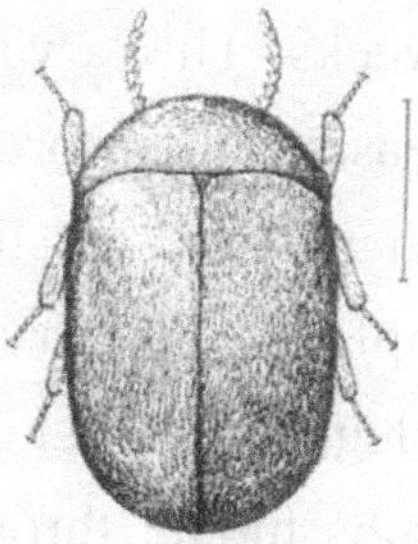

Order: Coleoptera

Family: Anobiidae

Scientific Name: Stegobium paniceum

Description:

Adult, light brown to red-brown, Oblong, moderately robust and if viewed from above, the head is not visible; hence, it appears hump-shaped. The body is covered with fine silky hair. Wings are uniquely punctate and distinctly striated. Antennae terminate in a lose 3 segmented club. They feed on all types of spices, milled products, biscuits, drugs, including opium, chilies, curry powder, turmeric, medicinal plants, books and manuscripts of Archeological importance. This is a primary pest destroying spices. Both adults and larvae are destructive; they can penetrate tin foils and lead sheets.

Biology:

Biologically, similar to the cigarette beetle. Adults are short-lived in the range of 2 to 4 weeks. The females lay about 75 to 100 eggs over a period of 8 to 10 days, depending upon the temperature. The normal larval period is about 3 to 5 weeks, pupates inside the cocoon, with a pupal duration 6 to 10 days. The entire life cycle is around 6 to 8 weeks. The larval and adult stages have intracellular symbiotic yeasts in their fore-guts in special pouches called mycetomes which make them independent of external sources of vitamin B complexes. The larva is Scarabaei form, less hairy, claws of thoracic legs are very long and slender, slightly curved. Active white grubs tunnel through the food materials. Pupa: having not mandibles formed inside the cocoon, which is adhered with food particles.

Habitat and Economic Importance:

Spice beetle feed on all types of food and spices. Signs of drugstore beetle are that the adults are seen buzzing around or resting on surfaces. Other signs include holes in packaging where the beetles have emerged and debris deposited next to infested items as the larvae feed. It can be difficult to find infestations in other areas. A flashlight will help with the inspection.

Common Site of Occurrence:

They infest majorly spices, milled products, Biscuits, drugs, curry powder, turmeric, medicinal plants, books and manuscripts. Distributed both in tropical and temperate region. Found in the large production and storage facilities of the above said products. Infested goods packed will also create problems in the retail outlets or in the residential units if the infested product is brought home. These insects are found where there is high relative humidity.

Management:

Controlling drugstore beetles begins with a thorough inspection. It is important to find everything that they have infested. Female sex pheromone traps are available for this beetle and can be used to identify areas of infestation. Heavily infested materials to be discarded. If there is light infestation, a heat treatment of 60 to 80 degree Celsius will kill all stages of the life cycle Books, Manuscripts, furniture and other infested might need to be fumigated. To prevent re-infestation, pyrethroid liquid or dust formulations can be applied to cracks, crevices, surfaces and other potential harborage areas. Product applications are supplemental to the cleanup of infested material and should never be used as the sole means of control

62.

Stable Fly

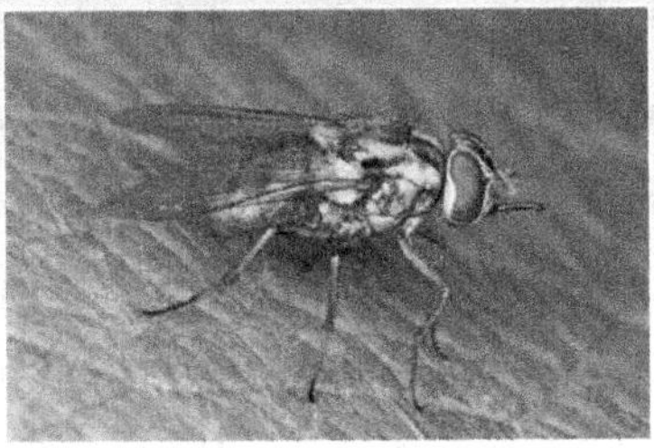

Order / Family: Diptera / Muscidae

***Scientific Name: Stomoxys calcitrans* (Linnaeus)**

Description:

The Stable Fly is a ¼ to 3/8 inch long, light grey fly, similar in appearance to House Fly. The Stable Fly has four dark stripes on the back of the thorax and has a light-coloured spot between the two centre stripes. It has a wider abdomen than House Fly and checker-board dark spots on the upper surface. The characteristic which separates the two is the bayonet-like mouthparts that project from the underside of the head. The mature larva is a spindle shaped maggot which is about 3/8 to ½ inch long. It has a slit-like, “S” – shaped respiratory opening at the end of its body.

Biology:

The female lays her eggs in rotting, fermenting, organic matter, including wet hay and straw, accumulated grass clippings, seaweed, manure/straw mixtures, and cannery wastes. The eggs hatch in one to three days and the larvae complete development in 11 - 30 days. Larvae moult three times before they pupate and remain in this stage for six to 20 days. Under optimum conditions, the life cycle takes 21 - 25 days.

Habitat and Economic Importance:

It is generally not a pest in houses but becoming a problem because new homes are being built in former agricultural areas. It can be annoying in outdoor areas around stables, livestock operations and on some beaches. The adults use its piercing-sucking mouthparts to feed on the blood of livestock, pets, and humans. It usually feeds on the lower part of the human body around the ankles. Its bite can be very painful.

Common Site of Occurrence:

These are predominantly filed insects or out of house insects. But the incidents of these insects being seen in households as well due to the increased urbanization in the agricultural fields as well. It normally lives in and around the live stock / poultry rearing places as they are blood feeders. It is not a insect of common in the inside building premises.

Management:

Control involves sanitation, mechanical control, and selective use of pesticides. Sanitation includes the complete removal and proper disposal of breeding materials essentials for larval development. Waste must be removed at least every two weeks to break the life cycles. Stall mucking should be spread in a thin layer over a field or disposed of at an approved landfill. Accumulations of vegetation, such as grass clippings on golf courses, sea weed on beaches, etc., should be removed. Screens should be properly maintained to prevent entry from structures. Yellow glue sticks are useful in areas where they are protected from dust and blowing debris. Adult flies can be controlled by applying wettable powder and micro encapsulated formulations to their resting sites.

63.

Stink Bug

Order: Hemiptera

Family: Pentatomidae

Scientific Name: Halyomopha halys
= Brown Marmorated stink bug

Description:

Pentatomidae is a family of insects belonging to the order Hemiptera, generally called Shield Bugs or Stink Bugs. As hemipterans, the Pentatomidae's have piercing sucking mouthparts, and most are phytophagous, including several species which are severe pests on agricultural crops. However, some species, particularly in the subfamily Asopinae, are predatory and may be considered beneficial. The name "stink bug" is specific to the Pentatomidae, and refers to their ability to release a pungent defensive spray when threatened, disturbed, or crushed. The composition of this spray may vary between species and even by sex or age but generally includes aldehydes and alkanes.

Biology:

Their combination of colour (brown, green, or metallic) and shape (oval, broad, and slightly convex) camouflages many of these insects as they sit on bark and foliage. The head and prothorax form a triangle that points forward. Stinkbugs generally range between 5 and 12 mm (0.2 and 0.5 inch) in length. The female deposits about 100 or more ornate barrel-shaped eggs in rows or clusters. Females of some species remain nearby and guard both eggs and newly hatched larvae. Eggs hatch within a week and the nymphs develop through 5 to 6 in stars and molts, becoming adult within about 5 weeks of hatching. The population is higher in the late summer and early winter period. In cooler climates, the adults hibernate during the winter.

Habitat and Economic Importance:

Stink Bugs may feed on plants, causing discolouration or spotting of the fruit, or prey on other insects. Stink Bugs can also be a significant annoyance to the homeowners congregating and flying around the exterior of homes buildings, entering through cracks and crevices. The bugs, however, are not known to cause any real harm to humans. Their name is derived from the odour that is eliminated from the scent glands when crushed; the glands are located on the dorsal surface or the abdomen and the underside of the thorax.

Common Site of Occurrence:

These are a pest of green or the garden or the vegetation area. They live and feed only on the vegetation seldom chances are there it to be seen inside the buildings unless it is bought inside along with the plants or the vegetation. They are common problem in the exterior house with ornamental plants or vegetation in the garden.

Management:

Like other pests, including ants and termites, Stink Bugs enter structures in larger numbers, making them difficult pests to control once inside. They can enter homes through the smallest openings, so it is important to repair or replace damaged screens on windows or doors. They are attracted to lights, so it is recommended to keep outdoor lighting to a minimum.

Recommended to use a vacuum cleaner for their removal, and also dispose of the vacuum bag immediately to prevent odour from permeating the area, as dead stink bugs leave a residue inside the bag that can stink up homes. Insecticide spray application of pyrethroid formulation can be applied to exterior cracks and crevices around doors, windows, vents and other potential entry points.

64.

Subterranean Termites

Order/Family: Isoptera/ Rhinotermitidae

Scientific Name:

Reticulitermes flavipes Kollar – Easter Subterranean Termite

Reticulitermes Hesperus Banks – Western Subterranean Termite

Description:

Termites are social insects that live in large colonies. These are three castes: reproductives, workers and soldiers. Termite antennae have bead-like segments. The winged reproductive (swarmer) have a pair of long wings (equal in size) attached to the last two thoracic segments. The wings are broken off after swarming. The abdomen is broadly joined at the thorax, unlike the narrow abdominal attachment found on ants. The winged reproductives are dark brown to almost black and about 3/8 inch long. The wings are brownish grey with few hairs and two dark veins on the leading edge. They have a very small pore on their heads. The soldiers are wingless with white bodies rectangular yellow-brown heads, which are two times longer than their width. And large mandibles which lack teeth.

Biology:

Subterranean termite colonies usually are located in the soil from which the workers build mud tubes to structural wood, where they then feed. Subterranean termite colonies are always connected to the soil and close to a moisture source.

Termites digest cellulose in wood with the aid of special organisms within their digestive system. The worker prefers to feed on fungus-infected wood but readily feed on undamaged wood as well. The foraging workers feed immature workers, reproductives, and soldiers with food materials from mouths and anuses.

A mature queen produces 5000 to 10000 eggs per year. An overage colony consists of 60000 to 250000 individuals, but colonies numbering in the millions are possible. A queen might live for up to 30 years and workers as long as 5 years.

Habitat and Economic Importance:

Subterranean colonies are established by winged reproductives, which usually appear in the spring. Swarms usually occur in the morning after a rain. A male and female that have swarmed from an established colony lose their wings and seek a dark cavity inside which they mate and raise the first group of workers. Both of these reproductives feed on wood, tend the eggs, and build the initial nest.

After the workers mature, they take over expanding the colony and feeding the reproductives. As the colony becomes larger, light coloured supplementary reproductives are produced to lay eggs which then become workers. The soldiers, which are also produced as the colony increases in size, are responsible for repelling invading ants and other predators.

Common Site of Occurrence:

These insects are called as soil insects as they breed, nest and form colonies only in the Soil. The reproductive alates will swarm during the first showers of monsoon and find a suitable place in the soil and start building their colony. These wood feeding insects so normally they were restricted to outdoors only feeding on dried leaves, barks, trees, however due to the usage of wood in the indoors and the urbanization going into greener areas the termites are gaining entry into the buildings through the gaps in the construction attached with the soil and gain entry. The presence of this can be unnoticed till we see a damage or a swarm of winged termites. These can infect any commercial, residential, manufacturing, institutions, etc any construction where there is wood source for it to feed.

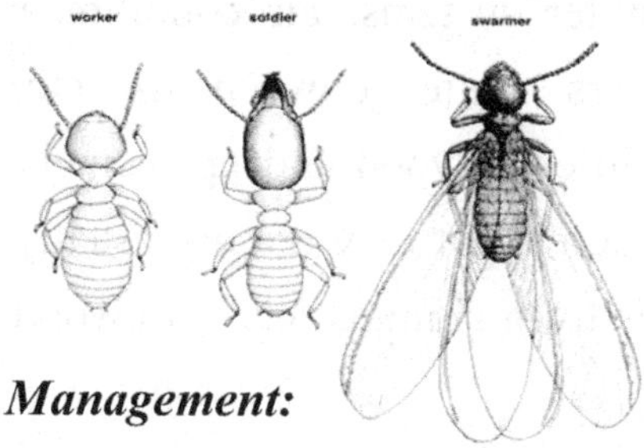

Management:

Subterranean termite infestations may go unnoticed until the winged reproductives "swarm" from or inside the structure. The presence of swarmer's is a good sign that a well-established colony is in the house and its immediate vicinity. Other evidence of infestation is wood damage, i.e., the spring or soft wood is eaten out of, leaving paper thin walls between the galleries, and the presence of mud tubes. Control begins with a thorough inspection to determine the extent of the infestation, the points of entry into the structure, moisture problems and other contributing factors, and the development of a plan to control. In addition, it is necessary to sound the structural timbers with a screwdriver or other shaper instrument in order to find galleries that have been excavated inside the wood.

The first step is to remove scarp wood, firewood, and any other wood materials that are in contact with the soil. There should be at least six inches between wood and/ or foam insulation and exterior soil grade and 18 inches between the soil in a crawl space and floor joists.

Water leaks and situations in which there is more wood and moisture content must be required. Drainage around the structure should be designed so that there is no accumulation of water near the foundation. Situations contributing to this condition include poor exterior soil grades, sprinkler systems, air conditioner condensate drains, lack of gutters, gutter downspouts, etc. Mechanical alterations, removal of infested wood, and replacement with treated wood are useful measures, but they are not usually successful in stopping the termites from gaining access to wood in the structure.

Galleries excavated in wood can be injected with aerosol residual insecticides to achieve temporary control of workers and winged reproductive in these localized areas. Treatment with EC formulations insecticide kills the termites and protects the treated wood from attack.

Detailed Ready Reckoner

Target Pest	Control Product / 's	Dosage
Coffee Bean Beetle; Cigarette Beetle; Spice Beetle / Drug Store Beetle; Confused flour Beetle / Rust Flour Beetle; Saw toothed grain Beetle / Merchant Grain Beetle; Khapra beetle; Flat Grain Beetle; Less Grain Borer; Grain Mites; Rice Moth / Indian Meal Moth; Almond Moth / Tropical Warehouse Moth; Angoumios Grain Moth / Paddy Moth; Granary Weevil / Wheat weevil; Rice Weevil; Maize Weevil; Cow pea weevil; Bean weevil; Furniture American	Preventive Measures as suggested and also the monitoring mechanism as suggested	NA
	Preventive Measures as suggested, Fumigation of stored products	Phosphine fumigation trials conducted across the country by CFTRI indicate a necessity of at least 500 ppm for the control of normal susceptible insect populations and more than >1000 ppm as the target concentrations for effective phosphine fumigation against phosphine-resistant insect populations in India.
	Prophylactic treatment on surfaces with Deltamethrin 2.5% WP	120 gms in 3 Ltrs of to be diluted and sprya to cover 100 sq mt area
	Primiphos Methyl 50% EC	30 ml in 3 ltrs of water to be diluted and spray to cover 100 sq mt area

Cockroach; Brown Banded Cockroach; German Cockorach; Oriental Cockroach; Smoky Brown Cockroach; Psocids; Book Lice; Spiders; Silverfish	Preventive Measures as suggested	NA
	Propoxur 20% EC	25 ml per ltr of water dilution covering 50 sq mt area
	Deltamethrin 2.5% SC	10 - 15 ml per ltr of water dilution covering 10 - 20 sq mt area
	Alphacypermethrin 10% SC	10 ml per ltr of water dilution covering 50 sq mt area
	Betacyfluthrin 2.45% SC	20 ml in 1 ltr of water dilution covering 20 sq mt area for porus and 40 sq mt area for non porous surface
	Cyfluthrin 5% EW	8 ml in 1 ltr of water dilution covering 50 ml spray solution per sq mt
	Fipronil 0.05% Gel	2 to 3 spots of 0.1gram per sq mt for American and Oreintal Cockroach; 1 to 2 spots of 0.1 gram per sq mt for German & Brown Banded Cockroach

	Cyphenothrin 5% EC	10 ml per ltr of water dilution covering 50 sq mt area for repeat application & 1st time application 20 ml per ltr of water
	Lamdacyhalothrin 10% WP	12.5 ml per ltr of water dilution covering 50 sq mt area
	Pyrethrum 2% EC	Product to be mixed in the ratio of 1:19 with Kerosene and used for very high infestations & resistant scenario
Brown Dog Tick; Cat Flea, Dog Flea; Dust Mites	Preventive Measures as suggested and Propoxur 20% EC	25 ml per ltr of water dilution covering 50 sq mt area on the surface of the wall, floor etc. but not on the animal
Bottle Fly / Blow Fly; House Fly; Stable Fly; Drain Moths; Fruit Fly	Preventive Measures as suggested	NA
	Propoxur 20% EC	25 ml per ltr of water dilution covering 50 sq mt area
	Deltamethrin 2.5% SC	10 - 15 ml per ltr of water dilution covering 10 - 20 sq mt area
	Alphacypermethrin 10% SC	10 ml per ltr of water dilution covering 50 sq mt area

	Betacyfluthrin 2.45% SC	20 ml in 1 ltr of water dilution covering 20 sq mt area for porus and 40 sq mt area for non porous surface
	Cyfluthrin 5% EW	8 ml in 1 ltr of water dilution covering 50 ml spray solution per sq mt
	Propoxur 2% RB	Place 20 to 25 gms per bait station as a loose bait & sprinkle few drops of water for flies to attract
	Imidacloprid 0.5% RB	Mix 25 g in 20ml and stirr it become paint like, using paint brush apply this mixture as a strip of not more than 10X30 cm
Fly Maggot Treatment	Diflubenzuron 25% WP	10 gm in 5 ltrs of water and sprayed for 10 sq mt area,
Fungus Gnats	Preventive Measures as suggested	NA
	Disinfectant	As per the manufacturers recommendation depending on the product usesd
	Residual spray of Insecticide	Dosage as per the Insecticide used

Aedes Mosquito; Culex Mosquito; Anopheles Mosquito;	Preventive Measures as suggested	NA
	Deltamethrin 2.5% SC & 2.5% WP	400 gm in 10 ltrs of water for coverage of 500 sq mt area for WP & 400 ml per 10 ltrs of water for coverage of 500 sq mt area for SC formulations
	Alphacypermethrin 10% SC & 5% WP	200 gm in 10 ltrs of water for coverage of 500 sq mt area for WP & 100 ml in 10 ltrs of water for SC formulation
	Betacyfluthrin 2.45% SC	20 ml in 1 ltr of water dilution covering 20 sq mt area for porus and 40 sq mt area for non porous surface
	Cyfluthrin 5% EW & 10% WP	125 gms in 10 ltrs of water for coverage of 500 sq mt area for WP & 125 ml per 10 ltrs of water for EW formulation
	Lamdacyhalothrin 10% WP & 10% CS	125 gms in 10 ltrs of water for coverage of 500 sq mt area for WP & 125 ml per 10 ltrs of water for EW formulation

	Bifenthrin 10% WP	125 gms in 10 ltrs of water for coverage of 500 sq mt area for WP & 125 ml per 10 ltrs of water for EW formulation
Mosquito Larva Treatment	Diflubenzuron 25% WP	2 gm in 2 ltrs of water and sprayed at the rate of 20 ml per sq mt coverage for clean water & 4 gm in 2 ltrs of water and sprayed at the rate of 20 ml per sq mt for polluted water
	Diflubenzuron 2% Granules	5 to 10kg per hectared for clean water; 10 to 20 kg per hectare for moderately polluted area, 20 to 50 kg per hectare for highly polluted area, 5 to 10 grams for 100 ltr water for ponds, stepwells etc
	Temephos 50% EC	2.5 to 5 ml per 10 ltrs of water for potable water, 7.5 to 10 ml per 10 ltr of water for moderately polluted water; 12.5 to 20 ml for heavily poluted water larvicidal treatement

	Temephos 1% Granules	5 to 10kg per hectared for clean water; 10 to 20 kg per hectare for moderately polluted area, 20 to 50 kg per hectare for highly polluted area, 5 to 10 grams for 100 ltr water for ponds, stepwells etc
	Bacillus Thurigenisis 1200 itu	20 ml per ltr of water covering 50 sq mt area
Thermal Fogging for Adult Mosquitoes for Outdoor	Deltamethrin 1.25 ULV	1:200 of the product to be mixed in diesel and used in a Thermal fogging machine
	Cyphenothrin 5% EC	8ml per ltr of diesel and used in thermal fogging machine.
	Pyrethrum Extract	1:19 of the product to be mixed in diesel and used in Thermal fogging machine
Cold Fogging for Adult Mosquitoes Indoor & Outdoor	Deltamethrin 2% EW	1:20 ratio to be mixed in water and Misting with ULV Machine or a Mist Spraying Machine to be used
	Cyphenothrin 5% EC	ULV Application of 1:8 times of Chemical to Water ratio & with Power Sprayer 1: 200 times chemical to water ratio

	Pyrethrum 2 EC	Product ratio of 1:19 to be mixed with water and used in a cold fogging machine or ULV machine
Norway Rats; House Mouse; Roof Rats; Bandicoot;	Preventive Measures and hygiene in terms of eliminating breeding / harborage area plays a key role	
	Bromodialone 0.005% Bait	Ready Bait Wax blocks cut into small blocks to be placed in a safe and secure place, far from reach of children and pets, preferably in Bait Stations
	Bromodialone 0.25% CB	A professional trained person to handle the product which can be mixed with any food bait material with 1:19 ratio of chemical & food material, which can be placed in bait stations for rodent control
	Brodifacum 0.005% Bait	Ready Bait Wax blocks cut into small blocks to be placed in a safe and secure place, far from reach of children and pets, preferably in Bait Stations

	Monitoring & Trapping Devices	Glue Pads, Snap Traps, Cage Traps, Bait Stations can be used to monitor the consumption pattern, species identification, probable area of probelem identifiation
Bats, House Sparrow, Pigeons	Preventive Measures is the most effective measures of control	
	Killing is prohibited in India	
	Scary Devices, Netting, Deterrent Glue, Dterrent Paste, Deterent Spikes are the methods used for detering the birds from the specific area of roosting / nesting	
Honey Bee	Killing of Honey bee is against the social structure being a beneficial insect; relocation of Honey Bee is the best practicse methods	

Other Insects like Field Cricket, Lady Bug, Social Wasp, Stink bug, Ear Wig, Scorpions, Millipede, Centipede, Caterpillar	These are rare insects of occurrence depending on the location and the habitat, it may also be let out as a part of nature and only if human interface is seen then the control meaures can b e adopted. Please refer to the specific insect details given in the book and adopt suitable measures	
Fire Ants, Pharaoh Ant, Litte Black Ant; Odorous House Ant; Black Carpenter Ant;	Preventive Measures is the most effective measures of control and the same to be followed as given in the book for better control methods	
	Propoxur 20% EC	To be used for spraying in cracks and crevices at the dose rate of 20 ml per ltr of water and sprayed for 50 sq mt area

	Imidacloprid 0.03% Gel	For Low infestation, One spot of 200 mg per sq mt and for high infestations, one spot of 300mg per Sq mt
Subterranean termites; Dampwood Termites; Drywood Termites;	Termite control is highly dependent on the prevention methods or control methods for which there is Pre Construction and Post Construction Termite Treatment. Please adhere to ghe BIS Specification IS 6313 for the treatment methods	
	Imidacloprid 30.5% SC	2.1 ml per ltr of water dilution for both pre construction and post construction termite treatement
	Fipronil 2.92 % EC	15 ml per ltr of water dilution for post construction anit termite treatment and 25 ml per ltr of water for pre construction anti-termite treatment. Can also be mixed with Kerosene for any wood treatment

	Bifenthrin 2.5% EC	20 ml per ltrs of water for Soil and Wall Treatment in the anti termite treatment
	Bifenthrin 2.5% EC	40 ml per ltr of Kerosene or Turbutaine oil for wood treatment against termites

Bibliography

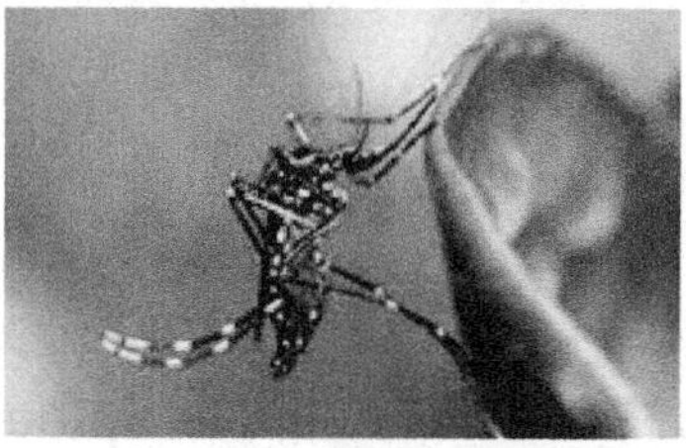

1. Aedes Mosquito

2. Almond Moth

3. American Cockroach

4. Angoumois Grain Moth

5. Anopheles Mosquito

6. Bats

7. Bed bug

8. Black Carpenter Ant

 9. Black Widow Spider	 13. Brown Reclusive Spider
 10.Bandicoot	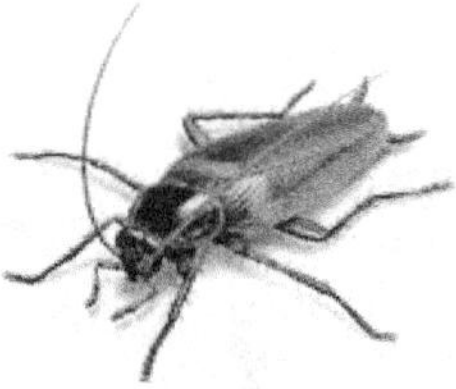 14. Brown Banded Cockroach
 11. Bottle/Blow Fly	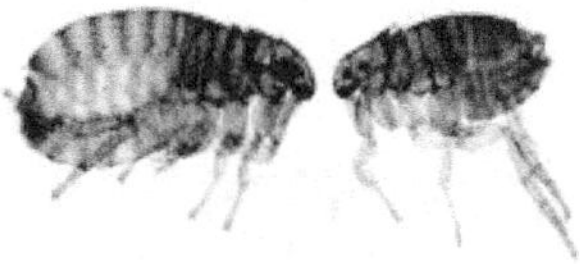 15. Cat/Dog Flea
 12. Brown dog tick	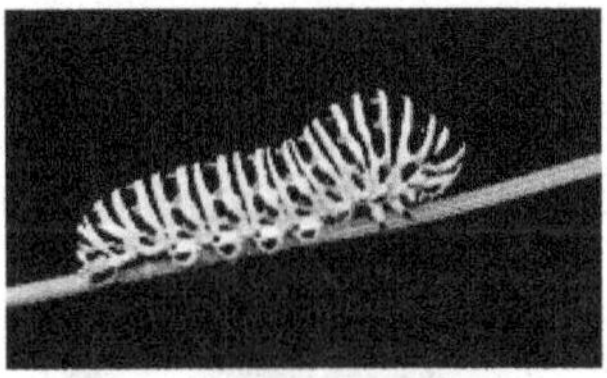 16. Caterpillar

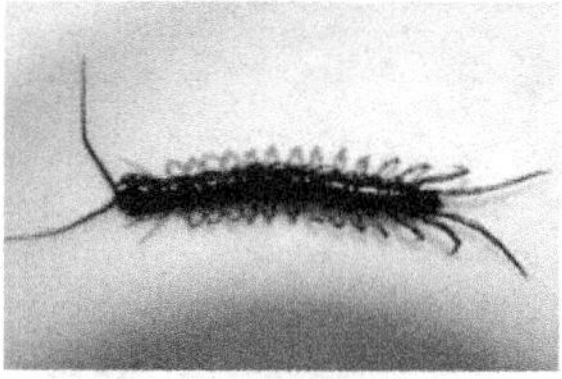

17. Centipede

18. Cigarette Beetle

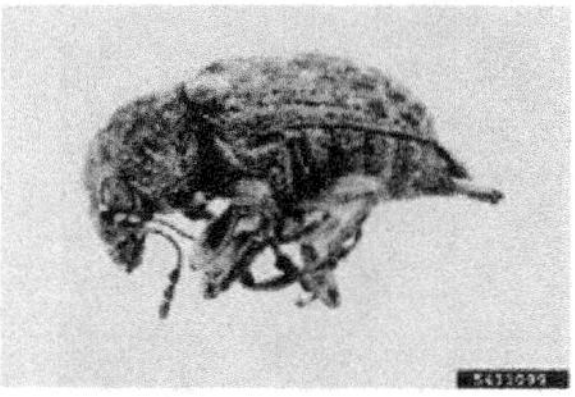

19. Coffee Bean Beetle

20. Common House Spider

21. Confused Flour Beetle

22. Cowpea Weevil

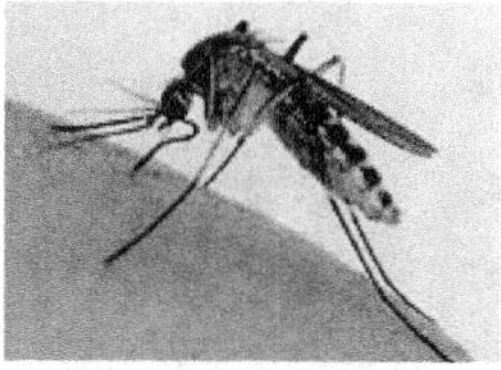

23. Culex Mosquito

24. Damp-wood Termite

25. Drain / Moth flies

26. Dry-wood Termites

27. Dust Mites

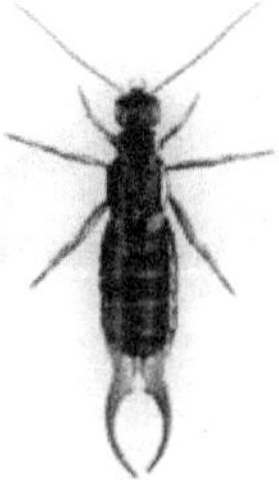

28. Earwig

29. Field Cricket

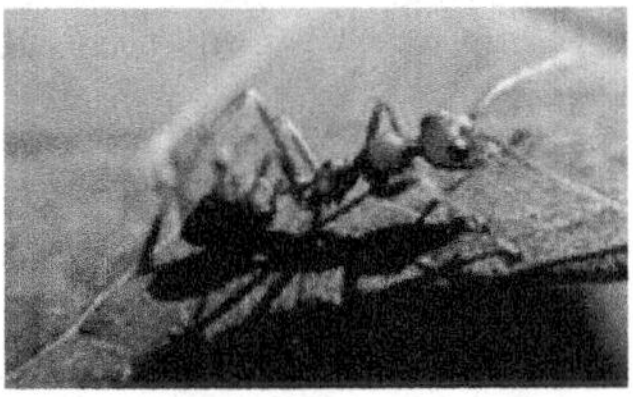

30. Fire/Red ants

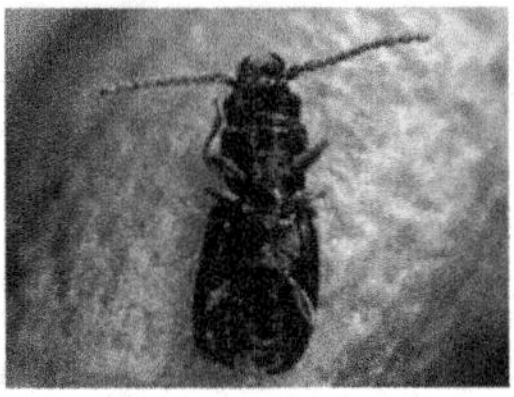

31. Flat Grain Beetles

32. Fruit fly

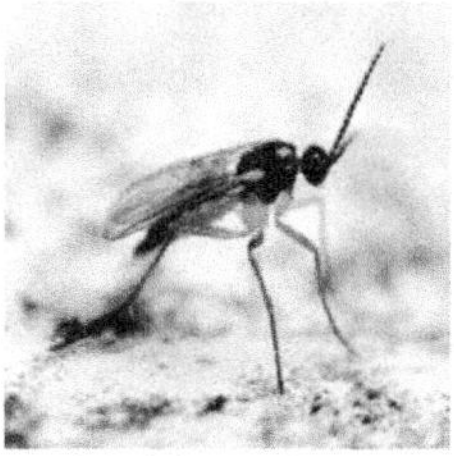

33. Fungus Gnats

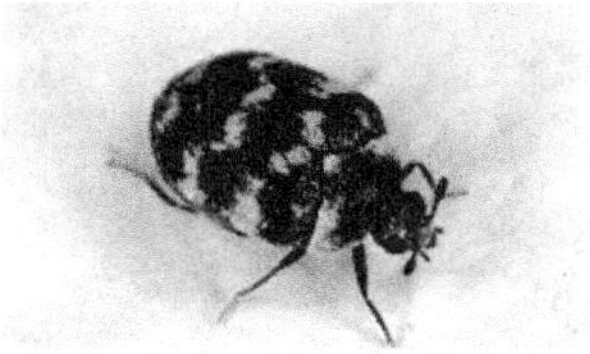

34. Furniture Carpet Beetle

35. German Cockroach

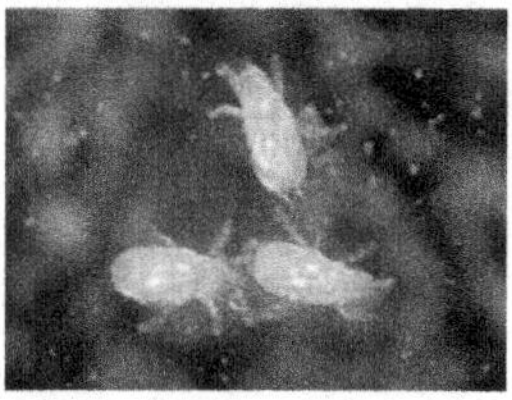

36. Grain mites

37. Granary Weevil

38. Honey Bee

39. House Fly

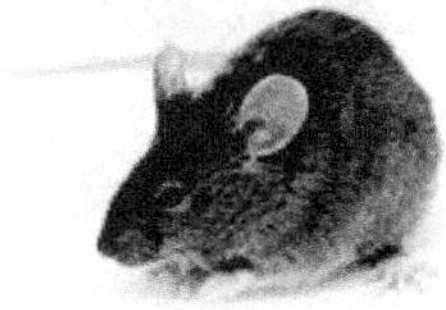

40. House Mouse

41. Sparrow

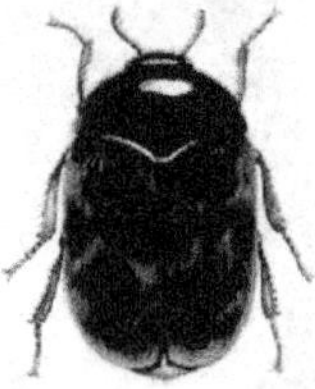

42. Khapra Beetle

43. Lady bug

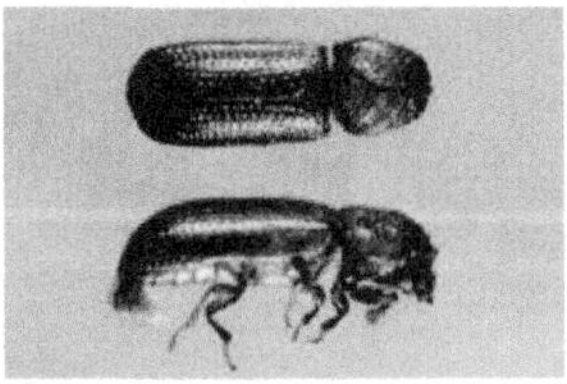

44. Lesser Grain Borer

45. Little Black ant

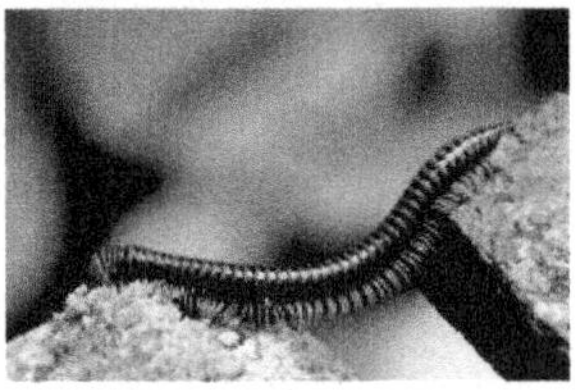

46. Millipede

47. Norway Rat

48. Odorous House Ant

49. Oriental Cockroach

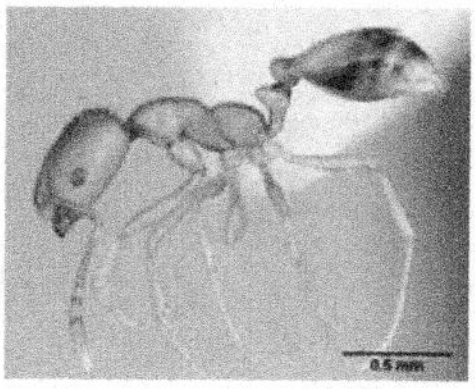

50. Pharaoh Ants

51. Pigeons

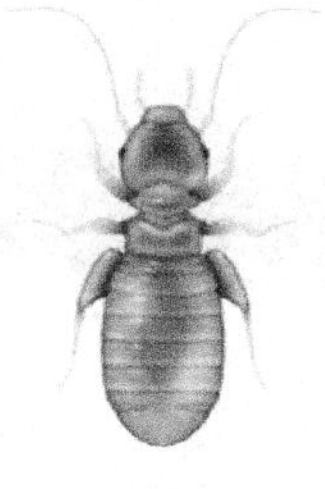

52. Psocids / Booklice

53. Indian Meal Moth

54. Rice Weevil

54. Maize Weevil

55. Roof Rat

56. Saw-toothed Grain beetle

57. Scorpions

58. Silverfish

59. Smoky Brown Cockroach

60. Social Wasp

61. Spice Beetle

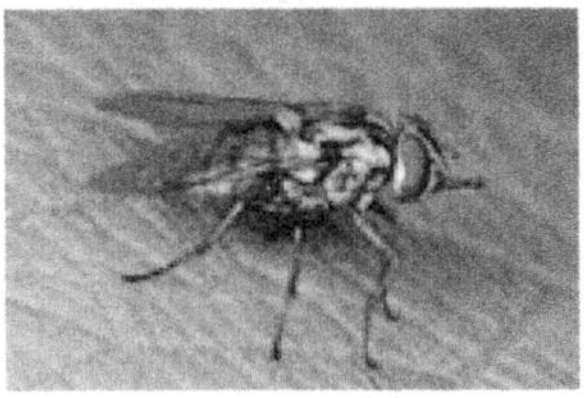

62. Stable fly

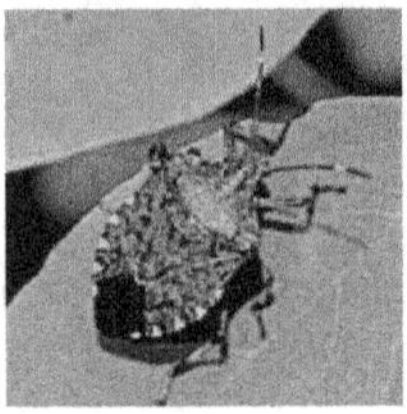

63. Stink bug

64. Subterranean Termites

REFERENCES

1. D.W. Hill, (1970) " Handling and storage of food grains in Tropical and subtropical Areas" FAO Agricultural development.
2. R.T. Cotton,(1963) "Pests of stored grain and grain products", Burgess Publ. Co., Minneapolis.
3. H.E. Hinton and A.S. Corbet,(1980) " Common Insect pests of stored food products."
4. J.W. Munro, (1966) "Pests of stored products", Hutchinson, London.
5. Tropical Development Research Institute(1984), " Insects and Arachnids of Tropical stored products --- their biology and identification' (A Training Manual).
6. O.L. Kurtz and K.L. Harris(1968) "Micro Analytical Entomology for Food Sanitation Control".
7. A.D. Imms(1951) " A General Text Book of Entomology", (8thEdn.).
8. V.B. Wigglesworth,(1965) " Principles of Insect Physiology", (Rev. Edn6th).
9. Pest Control Technology, Technicians Handbook (3rd edition)
10. A Concise Guide to INSECTS, by Patrick Hook
11. CFTRI Training on Fumigation, Prophylaxis & Pest Management Techniques for Stored Food Products

♦ ♦ ♦

Made in the USA
Monee, IL
04 November 2026

67381183R00154